BIOLOGICALLY-INSPIRED COLLABORATIVE COMPUTING

IFIP 20th World Computer Congress, Second IFIP TC 10 International Conference on Biologically-Inspired Collaborative Computing, September 8-9, 2008, Milano, Italy

Edited by

Mike Hinchey
Lero-the Irish Software Engineering Research Center
University of Limerick
Ireland

Anastasia Pagnoni
Università degli Studi di Milano
Italy

Franz J. Rammig
Universität Paderborn
Germany

Hartmut Schmeck
Karlsruhe Institute of Technology
Germany

T0137451

 Springer

Biologically-Inspired Collaborative Computing

Edited by Mike Hinchey, Anastasia Pagnoni, Franz J. Rammig
and Hartmut Schmeck

p. cm. (IFIP International Federation for Information Processing, a Springer Series in Computer Science)

ISSN: 1571-5736 / 1861-2288 (Internet)

ISBN: 978-1-4419-3502-1   eISBN: 978-0-387-09655-1

Printed on acid-free paper

9 8 7 6 5 4 3 2 1

springer.com

IFIP 2008 World Computer Congress (WCC'08)

Message from the Chairs

Every two years, the International Federation for Information Processing hosts a major event which showcases the scientific endeavours of its over one hundred Technical Committees and Working Groups. 2008 sees the 20th World Computer Congress (WCC 2008) take place for the first time in Italy, in Milan from 7-10 September 2008, at the MIC - Milano Convention Centre. The Congress is hosted by the Italian Computer Society, AICA, under the chairmanship of Giulio Occhini.

The Congress runs as a federation of co-located conferences offered by the different IFIP bodies, under the chairmanship of the scientific chair, Judith Bishop. For this Congress, we have a larger than usual number of thirteen conferences, ranging from Theoretical Computer Science, to Open Source Systems, to Entertainment Computing. Some of these are established conferences that run each year and some represent new, breaking areas of computing. Each conference had a call for papers, an International Programme Committee of experts and a thorough peer reviewed process. The Congress received 661 papers for the thirteen conferences, and selected 375 from those representing an acceptance rate of 56% (averaged over all conferences).

An innovative feature of WCC 2008 is the setting aside of two hours each day for cross-sessions relating to the integration of business and research, featuring the use of IT in Italian industry, sport, fashion and so on. This part is organized by Ivo De Lotto. The Congress will be opened by representatives from government bodies and Societies associated with IT in Italy.

This volume is one of fourteen volumes associated with the scientific conferences and the industry sessions. Each covers a specific topic and separately or together they form a valuable record of the state of computing research in the world in 2008. Each volume was prepared for publication in the Springer IFIP Series by the conference's volume editors. The overall Chair for all the volumes published for the Congress is John Impagliazzo.

For full details on the Congress, refer to the webpage http://www.wcc2008.org.

Judith Bishop, South Africa, Co-Chair, International Program Committee
Ivo De Lotto, Italy, Co-Chair, International Program Committee
Giulio Occhini, Italy, Chair, Organizing Committee
John Impagliazzo, United States, Publications Chair

WCC 2008 Scientific Conferences

TC12	AI	Artificial Intelligence 2008
TC10	BICC	Biologically Inspired Cooperative Computing
WG 5.4	CAI	Computer-Aided Innovation (Topical Session)
WG 10.2	DIPES	Distributed and Parallel Embedded Systems
TC14	ECS	Entertainment Computing Symposium
TC3	ED_L2L	Learning to Live in the Knowledge Society
WG 9.7 TC3	HCE3	History of Computing and Education 3
TC13	HCI	Human Computer Interaction
TC8	ISREP	Information Systems Research, Education and Practice
WG 12.6	KMIA	Knowledge Management in Action
TC2 WG 2.13	OSS	Open Source Systems
TC11	IFIP SEC	Information Security Conference
TC1	TCS	Theoretical Computer Science

IFIP

- is the leading multinational, apolitical organization in Information and Communications Technologies and Sciences
- is recognized by United Nations and other world bodies
- represents IT Societies from 56 countries or regions, covering all 5 continents with a total membership of over half a million
- links more than 3500 scientists from Academia and Industry, organized in more than 101 Working Groups reporting to 13 Technical Committees
- sponsors 100 conferences yearly providing unparalleled coverage from theoretical informatics to the relationship between informatics and society including hardware and software technologies, and networked information systems

*Details of the IFIP Technical Committees and Working Groups
can be found on the website at http://www.ifip.org.*

Preface

"Look deep into nature and you will understand everything better." advised Albert Einstein.

In recent years, the research communities in Computer Science, Engineering, and other disciplines have taken this message to heart, and a relatively new field of "biologically-inspired computing" has been born. Inspiration is being drawn from nature, from the behaviors of colonies of ants, of swarms of bees and even the human body. This new paradigm in computing takes many simple autonomous objects or agents and lets them jointly perform a complex task, without having the need for centralized control. In this paradigm, these simple objects interact locally with their environment using simple rules. Applications include optimization algorithms, communications networks, scheduling and decision making, supply-chain management, and robotics, to name just a few.

There are many disciplines involved in making such systems work: from artificial intelligence to energy aware systems. Often these disciplines have their own field of focus, have their own conferences, or only deal with specialized sub-problems (e.g. swarm intelligence, biologically inspired computation, sensor networks). The Second IFIP Conference on Biologically-Inspired Collaborative Computing aims to bridge this separation of the scientific community and bring together researchers in the fields of Organic Computing, Autonomic Computing, Self-Organizing Systems, Pervasive Computing and related areas.

We are very pleased to have two very important keynote presentations:

- *Swarm Robotics: The Coordination of Robots via Swarm Intelligence Principles* by Marco Dorigo (Université Libre de Bruxelles, Belgium), of which an abstract is included in this volume.
- *Immuno-engineering* by Jon Timmis and his collaborators at University of York, UK (full paper included in this volume).

The contributions to the program of this conference have been selected from submissions originating from North and South America, Asia, Europe and Australia. We would like to thank the members of the program committee for the careful reviewing of all submissions, which formed the basis for selecting this attractive program. We are grateful to IFIP and in particular IFIP TC-10 for their support.

We welcome all participants of this Second IFIP Conference on Biologically-Inspired Collaborative Computing—BICC 2008—and look forward to an inspiring series of talks and discussions, part of a range of excellent conferences in the IFIP World Computer Conference 2008.

Franz J.Rammig (Germany)   *Hartmut Schmeck* (Germany)
Anastasia Pagnoni (Italy)   *Mike Hinchey* (Ireland)
(Conference Co-Chairs)   (Program Co-Chairs)

Program Committee

Jürgen Branke, University of Karlsruhe, Germany
Sven Brueckner, New Vectors LLC, USA
Yuan-Shun Dai, University of Tennessee at Knoxville, USA
Giovanna Di Marzo Serugendo, University of London, UK
Marco Dorigo, IRIDIA, Université Libre de Bruxelles, Belgium
Luca Maria Gambardella, IDSIA, Switzerland
Xiaodong Li, Royal Melbourne Institute of Technology, Australia
Peter Lindsay, University of Queensland, Australia
Tiziana Margaria, University of Potsdam, Germany
Eliane Martins, UNICAMP, Brazil
Christian Müller-Schloer, Universität Hannover, Germany
Roy A. Maxion, Carnegie Mellon University, USA
Takashi Nanya, RCAST, University of Tokyo, Japan
Bernhard Nebel, Albert-Ludwigs-Universität Freiburg, Germany
Jochen Pfalzgraf, Universität Salzburg, Austria
Daniel Polani, University of Hertfordshire, UK
Ricardo Reis, Universidade Federal do Rio Grande do Sul, Brazil
Richard D. Schlichting, AT&T Labs, USA
Bernhard Sendhoff, Honda Research Institute, Germany
Henk Sips, Delft University of Technology, The Netherlands
Leslie Smith, University of Stirling, UK
Albert Y. Zomaya, University of Sydney, Australia

Additional Reviewers
Marco Bakera, University of Dortmund, Germany
Sven Jörges, University of Dortmund, Germany
Georg Jung, University of Potsdam, Germany
Lukas König, University of Karlsruhe, Germany
Carlo Pinciroli, Université Libre de Bruxelles, Belgium
Clemens Renner, University of Dortmund, Germany

Contents

Swarm Robotics: The Coordination of Robots via Swarm Intelligence Principles

Marco Dorigo

IRIDIA, CoDE, Université Libre de Bruxelles

Abstract: Swarm intelligence is the discipline that deals with natural and artificial systems composed of many individuals that coordinate using decentralized control and self-organization. The discipline focuses on the collective behaviors that result from the local interactions of the individuals with each other and with their environment. Examples of systems studied by swarm intelligence are colonies of ants and termites, schools of fish, and flocks of birds. Some human artifacts also fall into the domain of swarm intelligence. Examples are, among others, some multi-robot systems and certain computer programs tackling optimization and data analysis problems.

After a short introduction to swarm intelligence, in my presentation I will focus on recent work in swarm robotics, that is, the application of swarm intelligence principles to the control of swarms of cooperating robots. In particular, I will present results of the swarm-bot experiment. A swarm-bot is an artifact composed of a swarm of assembled s-bots, mobile robots capable of connecting to, and disconnecting from, each other. In the swarm-bot form, the s-bots are attached to each other and, when needed, become a single robotic system that can move and change its shape. S-bots have relatively simple sensors and motors and limited computational capabilities. A swarm-bot can solve problems that cannot be solved by s-bots alone.

In the talk, I will briefly describe the s-bots hardware and the methodology followed to develop algorithms for their control. Then I will focus on the capabilities of the swarm-bot robotic system by showing video recordings of some of the many experiments we performed to study coordinated movement, path formation, self-assembly, collective transport, shape formation, and other collective behaviors.

Please use the following format when citing this chapter:

Dorigo, M., 2008, in IFIP International Federation for Information Processing, Volume 268; *Biologically-Inspired Collaborative Computing*; Mike Hinchey, Anastasia Pagnoni, Franz J. Rammig, Hartmut Schmeck; (Boston: Springer), p. 1.

Immuno-engineering

Jon Timmis[1], Emma Hart[3], Andy Hone[4], Mark Neal[5] Adrian Robins[6], Susan Stepney[1], and Andy Tyrrell[2]

Abstract In this position paper, we outline a vision for a new type of engineering: immuno-engineering, that can be used for the development of biologically grounded and theoretically understood Artificial Immune Systems (AIS). We argue that, like many bio-inspired paradigms, AIS have drifted somewhat away from the source of inspiration. We also argue that through an interdisciplinary approach, it is possible to exploit the underlying biology for computation in a way that, as yet, has not been achieved. Immuno-engineering will not only allow for the potential development of more powerful AIS, but allow for feed back to biology from computation.

1 Introduction

Advances in technology today enable the construction of complex autonomous systems, which can range in size from a robot, perhaps containing tens of simple devices, to mobile, ad–hoc networks containing thousands of such devices. At both extremes, such systems consist of unreliable heterogeneous sensors and actuators which must make decisions across multiple timescales in unpredictable, and potentially hostile, dynamic environments in order to maintain their integrity and achieve their desired functionality. Current technology allows us to hard-wire responses to foreseeable situations; a considerable void is still to be crossed however to achieve systems which adapt continuously and autonomously to their environments and exhibit what is becoming known as self-CHOP characteristics; self-configure, self-heal, self-optimize and self-protect. A paradigm shift in engineering is required to address this; we propose that a new discipline that will allow for the construction

Department of Computer Science, University of York, York. UK e-mail: jtimmis@cs.york.ac.uk · Department of Electronics, University of York, York. UK · School of Computing, Napier University, Edinburgh, UK · Institute of Mathematics, Statistics and Actuarial Science, University of Kent, Canterbury, Kent. UK · Department of Computer Science, Aberystwyth University, Aberystwyth, UK · University of Nottingham Medical School, Nottingham, UK

Please use the following format when citing this chapter:

Timmis, J., Hart, E., Hone, A., Neal, M., Robins, A., Stepney, S. and Tyrrell, A., 2008, in IFIP International Federation for Information Processing, Volume 268; *Biologically-Inspired Collaborative Computing*; Mike Hinchey, Anastasia Pagnoni, Franz J. Rammig, Hartmut Schmeck; (Boston: Springer), pp. 3–17.

of engineered artefacts that are fit for purpose in the same way as their biological counterparts needs to be developed.

Our long-term aim is to develop the foundations for a new kind of engineering – *immuno-engineering* – exploiting principles derived from the human immune system to enable the engineering of robust complex artefacts.

In this position paper we outline the concept of Immuno-engineering and discuss it's motivation in the context of current Artificial Immune System (AIS) research, and we hint at the way in which such a discipline may be developed.

We propose a bottom-up approach to the engineering of such systems, which will result in a set of immuno-engineering principles; these can be generalised to the future development of a wide range of bio-inspired, autonomic systems. This is achieved via an interdisciplinary approach which cuts across immunology, mathematics, computer science and engineering. In a recent paper [1] we discuss the interdisciplinary nature of AIS research. Our vision is inspired by recent work in immunology which attempts to reposition the immune system away from a pure defence mechanism to a complex, self-organising *computational system*, which computes the state of the body and then responds to it in order to achieve host maintenance and protection [2]. Autonomic systems which operate in a dynamic and information rich environment need to compute their state and then respond in an analogous way if they are to remain operational in order to continuously deliver the services expected of them.

2 Exploiting Immunology for Computation

Artificial Immune Systems (AIS) [3] is a diverse area of research that attempts to bridge the divide between immunology and engineering and are developed through the application of techniques such as mathematical and computational modeling of immunology, abstraction from those models into algorithm (and system) design and implementation in the context of engineering. Many early attempts to apply immunological inspiration to engineering began with efforts to mimic the perceived role of the natural immune system as a mechanism for identifying and then eliminating harmful pathogens from the body in a computer intrusion detection system [4]. However, work previous to this explored the immune system for inspiration in fault diagnosis [5] and control [6]. These investigations sparked a host of attempts to apply aspects of immunology to a wider range of engineering problems, and the reader is referred to the International Conference on Artificial Immune Systems (ICARIS) for a comprehensive collection of papers [7, 8, 9, 10, 11, 12]. Over recent years there have been a number of review papers written on AIS with the first being [13] followed by a series of others that either review AIS in general, for example, [14, 15, 16, 17, 18], or more specific aspects of AIS such as data mining [19], network security [20], applications of AIS [21], theoretical aspects [22] and modelling in AIS [23].

However, despite the many successes of the immune inspired approach we claim that the real potential of the approach has yet to be met [18]. We claim that this results from two limitations in the approach taken. Firstly, all of these applications have cherry picked one (or occasionally a few) features of the vertebrate immune system and attempted to apply them in isolation. Thus, we observe algorithms based on clonal selection e.g. [24], on negative selection [25], on idiotypic networks [26] and dendritic cells [27] with many recent developments in AIS being based around one of these four types of algorithms. Moreover, almost without exception there has been a tendency to exploit only the adaptive component of the vertebrate immune system. It is clear from immunological studies that the innate and adaptive components operate in tandem, and furthermore, regulate each other's effects. Therefore, by selecting only individual components of a complex, interacting system, a huge opportunity to exploit the true potential of the metaphor is being missed. Secondly, the focus on individual applications has followed an approach common to much bio-inspired research: an algorithm is designed and tuned empirically to a particular problem, thereby making it difficult to generalise any principles applicable to other applications. The EPSRC "Danger Theory" project[1], the outlines of which were proposed in [28] was the first attempt to combine current immunological experimentation with computational research. However, even it has focussed on a single application (intrusion detection [20]) and a single aspect of the immune system (danger theory) [27, 29, 30, 31]. Whilst this research has provided significant developments in the area of intrusion detection, we feel that for the area of AIS to progress we need to find more general principles that are applicable in a range of application areas. It should be noted that it is not feasible to capture the whole immune system in a single application, the sheer complexity would be overwhelming, however a focus on higher-level key properties, such as multiple-timescales of response, of the innate and adaptive components, may prove useful in their generic applicability to engineering.

Based on these observations, we feel the time is ripe for a *step change* in the development of AIS, through a principled engineering approach. We now discuss such an approach.

3 Defining Immuno-engineering

We follow Orosz's definitions of 'immuno-ecology' and 'immuno-informatics' [32]:

immuno-ecology   : "the study of immunological principles that permit effective immunological function within the context of the immensely complex immunological network . . . the principles serve mainly to provide an infrastructure for the immune system."

---

[1] http://www.dangertheory.com/

immuno-informatics   : "the study of the immune system as a cognitive, decision-making device ... addresses mechanisms by which the immune system converts stimuli into information, how it processes and communicates that information, and how the information is used to promote an effective immuno-ecology ... how the immune system generates, posts, processes, and stores information about itself and its environment"

and so we now define:

immuno-engineering: the abstraction of immuno-ecological and immuno-informatics principles, and their adaptation and application to engineered artefacts (comprising hardware and software), so as to provide these artefacts with properties analogous to those provided to organisms by their natural immune systems.

Immuno-engineering takes into account the differences between artificial systems and biological systems: for example, the different numbers, kinds, and rates of signals that need to be monitored and processed; the different kinds of decisions that need to be made; the different effectors available to support and implement those decisions; and the different constraints of embodiment, either physically or virtually engineered. For example, Orosz [32] hypothesises that the major design features of the biological immune system that provides speed, flexibility and multiple response options rely on a parallel-processing system which has 'wasteful' use of resources, countless back-up systems, and requires the ability to immediately and continuously monitor physical sites. This is of enormous consequence to the engineer, who is constrained by processing speeds, communication overheads, and physical resources, and furthermore hindered by hardware requirements such as transmitting signals from sensors, but who can freely make numerous copies of software agents, subject only to storage constraints.

3.1 A Conceptual Framework for the Development of Immuno-engineering

In their paper, Stepney *et al.* [33] propose that bio-inspired algorithms, such as AIS, are best developed in a more principled way than was currently being undertaken in the literature. To clarify, the authors suggested that many AIS recently developed had drifted away from the immunological inspiration that had fueled their development and that AIS practitioners were failing to capture the complexity and richness that the immune system offers. In order to remedy this, the authors suggest a conceptual framework for developing bio-inspired algorithms within a more principled framework that attempts to capture biological richness and complexity but, at the same time, appreciate the need for sound engineered systems that need to work. This should avoid the "reasoning by metaphor" approach often seen in bio-inspired computing whereby algorithms are just a weak analogy of the process on which they are based, being developed directly from (often naive) biological models and observations. One of the main problems involved in designing bio-inspired algorithms is deciding which aspects of the biology are necessary to generate the required behaviour and which aspects are surplus to requirements. Thus, the conceptual framework takes an interdisciplinary approach, involving the design of AIS through a series of observational and modelling stages in order to identify the key characteristics of the immunological process on which the AIS will be based. The first stage of the conceptual framework, as outlined in figure 1, aims to probe the biology, utilising biological observations and experiments to provide a partial view of the biological system from which inspiration is being taken. This view is used to build abstract models of the biology. These models can be both mathematical and computational, and are open to validation techniques not available to the actual biological system. From the execution of the models and their validation, insight can be gained into the underlying biological process. It is this insight that leads to the construction of the bio-inspired algorithms. This whole process is iterative, and can also lead to the construction of computational frameworks that provide a suitable structure for specific application-oriented algorithms to be designed from.

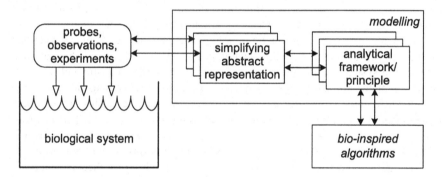

Fig. 1 The Conceptual Framework [33]. This can be seen as a methodology to develop novel AIS allowing true interaction between disciples where all can benefit, and, a way of thinking about the scope of AIS and how that has broadened over the years once again

As noted by Stepney *et al.* [33] each step in the standard conceptual framework is biased, be it modelling some particular biology mechanism or designing an algorithm for which there is an intended end product or specific concept. The first instantiations of the conceptual framework will produce models specific to certain biological systems and algorithms for solutions to specific problems. One could attempt to produce a computational framework based on some biology without a particular end algorithm/application in mind, that is examining biology and hoping to come across something applicable to a generic computational problem. This, however, would seem to be a very difficult task and one has to ground the development of AIS in some form of application at some point. Therefore, it is far easier to orient these steps toward some particular problem giving necessary focus to the modelling work [34].

4 Towards Immuno-engineering

From an *engineering* perspective, there is a need to design, develop and implement *design libraries* derived from the immuno-engineering principles, tested in diverse practical exemplars ranging from a self-contained and self-maintaining piece of hardware, such as a network switch that reduces downtime and error-propagation on the internet, to a ubiquitous sensing system with reliable message passing that could be embedded into buildings resulting in a system that could rapidly detect and localise survivors following building collapse. From a *biological* perspective, by focussing on the immune system as a computational system we will deliver a framework in which it is possible to reframe experimental immunological data and ask new experimental questions. For example, we might ask how the state of a developing tumor can influence the state of the tissues and therefore how can we induce the immune system to compute this state as abnormal. Recent work in [2] has developed the notion of the "computation of the state" of the immune system: as stated earlier, an immuno-engineering approach should take this into account. Such an interdisciplinary endeavor will thus potentially impact on the understanding of disorders of immune activity such as autoimmune diseases and cancer.

To provide this bridge between immunology and engineering, there is a potential need to utilise state-based modelling techniques which fit well with a computational view of the systems. On the one hand, these will provide a realistic and intuitive environment for immunologists to complement traditional mathematic modelling such as differential equations and, on the other, they can be readily transformed into engineering solutions. One such approach is the π-calculus [35]. This is a formal language used to specify concurrent computational systems. Its defining feature that sets it apart from other process calculi is the possibility of expressing mobility. This allows processes to "move" by dynamically changing their channels of communication with other processes, thus one can model networks that reconfigure themselves. The π-calculus allows composition, choice, and restriction of processes which communicate on potentially private complementary channels. There is a growing similarity between the parallelism and complexity of computer systems today and biological systems. As noted by [36] computational analysis tools such as the π-calculus are just as applicable to biology as they are to computing.

4.1 What do we need to do for Immuno-engineering?

The properties we wish to endow on engineered systems are currently exhibited only by those complex living systems whose immune systems comprise an *innate* component which endows the host with rapid *pre-programmed* responses *and* an *adaptive* component which is capable of learning through experience. Much of the desired functionality of the system arises from the interplay between these subsystems and the regulatory effect they have on each other. Together, these operate over multiple timescales, from seconds to the entire lifetime of the organism. Therefore, the modelling of both innate and adaptive components, paying particular attention to the interface between them, enables us to push the boundaries of biologically inspired computing and engineering.

In order to achieve our aim of laying the foundations for immuno-engineering, a number of key objectives need to be achieved:

1. **derive** mathematical models of the interplay between the innate and adaptive immune systems
2. **develop** and **verify** computational models that capture the interplay of innate and adaptive immunity
3. **implement** an immuno-engineering design and implementation library
4. **develop** and **assess** immuno-engineering insights to inform modulation of the natural immune system
5. **deploy** and **evaluate** the immuno-engineering library in a diverse set of case studies

5 Instantiating Immuno-engineering

In order to develop the Immuno-engineering approach a combination of the conceptual framework [33] and the problem-oriented perspective [34] can be adopted and requires interactions between computation, mathematical analysis, practical implementation, and biological experimentation. These will be rooted in the conceptual framework [33], which formulates the principled abstraction of bio-inspired algorithms through a process of mathematical modelling, computational modelling, and algorithm development for application domains, and makes use of the problem-oriented perspective [34] through the use of case studies to develop and refine the Immuno-engineering libraries.

Work should be based on a combination of mathematical and computational modelling, which leads to the development of an immuno-engineering library. This library should be tested on a number of carefully selected, realistic and diverse case studies that exhibit a broad and diverse spectrum of engineering features, thus allowing for the refinement of the approach. The library should thus be exercised in various different forms (in hardware or software, in open or closed environments), and therefore be tested and evaluated. Such suitable case studies might include webmining, condition-monitoring and distributed sensing, in line with suggestions for future applications of AIS [21] which emphasises the notion of dynamic, life-long learning and homeostasis. In addition, elements of our immuno-engineering library should be used to computationally model aspects of the human immune system, which will help inform *in vitro* experiments: hence, all disciplines in this endeavor will benefit from the whole approach.

Figure 2 illustrates the interdisciplinary nature of the work. In order to develop the Immuno-engineering approach we would advocate focussing on the development of mathematical models of interactions between the innate and adaptive

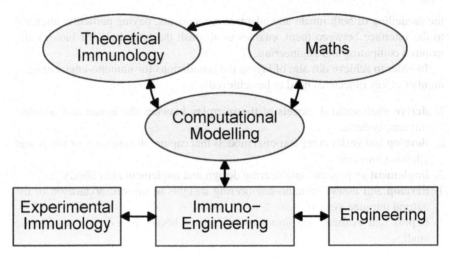

Fig. 2 Interactions between disciplines that leads to the development of immuno-engineering which itself acts as the bridge between experimental immunology and engineering

immune systems, capturing the essentials of immuno-ecology (the interaction between theoretical immunology and mathematics). Once mathematical models have been developed, we can proceed to develop computational models from them: this will lay the foundation for developing immuno-engineering and develop an *immuno-engineering library*. This library will *bridge the gap* between experimental immunology and engineering, thus breaking the mould of typical biologically inspired computing, which simply jumps from simplistic views of biological systems straight to simplistic engineered solutions, in line with the approach advocated in [33]. With the adoption of this method, it should be possible to generate a set of mathematically sound, biologically grounded techniques applicable to engineering. In addition, the library will drive further investigations into experimental immunology.

5.1 Modelling and Immuno-Ecology

Modelling provides us with a fundamental insight into the workings of the immune system. Inherent in our proposal is the desire to gain a detailed understanding of immunological principles which will ultimately lead to the development of the immuno-engineering library. Modelling affords us the opportunity to investigate a complex system from different perspectives: from the level of individual components (molecules and cells), to the level of populations of cells, to an overall systems level. For the modelling, a wide variety of options are open, all with their own advantages and disadvantages [23] such as dynamical systems, optimal control theory, information and coding, probability, stochastic π-calculus and complex network theory.

5.1.1 Modelling of Information Processing

A central part of the interaction between the innate and adaptive immune system is the process by which antigen is recognised. In order to derive mathematical models of *antigen processing*, one could adopt an information-theoretic approach to the identification of antigens, regarding them as salient chunks of data to be processed by the immune system. Antigens usually take the form of proteins which are recognized as being "foreign" by the immune system, and, for this recognition to occur, suitable features of the structure and composition of the protein must first be isolated. The latter is achieved by immune cell receptors which bind to specific chunks of the antigen. Regarding proteins as streams of data will enable the use of information theory to model the processing of such data.

Having obtained an abstract representation of antigen processing, it should then be possible to produce dynamic models of the *regulation* of receptor-bearing agents, based on, for example, optimal control theory. These would involve both continuous and discrete dynamics (differential and difference equations respectively). By performing analytical studies of the mathematical models, their asymptotic behaviour can be determined. These dynamical systems will also be converted into numerical simulations suitable for developing computational models, which will provide numerical predictions that can be compared with the results of the development of the algorithms.

5.1.2 Modelling of Network Topologies

It is possible to examine the way that the overall response is mediated by the network of signals created by cytokines. A sensible approach would be to extract some generic topological features by focusing on particular small subsets of the immune network, chosen for the relevance to the desired engineering properties. In order to do this one could look towards complex network theory, using as a starting point the models of Barabasi, Watts & Strogatz, and extending this with ideas of Alon on network motifs for biological systems [37]. Specific subsets of immunological networks to be examined include: self regulation and switching between immune cell responses such as T-helper cells e.g. Th1/Th2; innate modulation of adaptive immune response via the complement system (which helps clear pathogens from an organism); mutually inhibitory effects of cytokines such as IL4 and IFγ upon one another.

Network models of specific subsystems such as these will be developed, and can then be tested against experiments, leading to further refinement of these models. Through the biological experiments it should be possible to verify features of the local topology of subsystems within immune networks. It would be possible to investigate how the particular architecture of these networks affects the robustness of their response, allowing the selection of suitable network motifs to be incorporated into the development of the Immuno-engineering library.

5.2 *Immuno-engineering Library*

Adopting a problem-oriented approach [34] it should be possible to determine the immuno-requirements of the case studies helping to drive the research and determine the relevant Immuno-Ecology and Immuno-Informatics principles. From these bio-specific principles, it will be possible to develop abstract descriptions in the form of UML (statecharts, sequence diagrams) and develop design patterns [38] that capture salient properties and encapsulate constraints. Their properties could be analysed to ensure that they have not lost the desired immuno-properties during abstraction. This should be iterated as appropriate, incorporating extra components as discovered through the modelling work and the case studies. The output of this task would comprise generative pattern languages of immuno-engineering analysis and design: tools that aid a system designer in analysing a specific application, and in applying immuno-engineering principles and concepts to its particular sensors, tasks, and embodiment that respects the biological underpinnings.

The abstractions and pattern languages that have now been developed will provide the foundation for building a software library for developing immuno-engineered systems. This would result in an implementation library in the form of architectures and algorithms that can be instantiated in the case studies. One of the fundamental properties of the immune system is that it is a highly parallel system of communicating agents. Therefore, consideration should be given to potential parallel processing architectures, platforms and programming languages and determine which will allow the immuno-engineering properties to be implemented effectively, and what physical constraints they will impose. For example, technologies such as VHDL, JSCP and occam-π allow for the development of truly parallel software systems and allow for a natural mapping from the stochastic-π calculus.

5.3 *Experimental Immunology*

Central to the concept of immuno-engineering is the ability to employ immuno-engineering principles in the context of actual experimental immunology. Immuno-engineering principles and mathematical models developed through this process should direct experimental work. Indeed, one of the major motivations for the use of modelling in experimental work is to make use of the predictive nature of the models and to tie them closely to experimental work: otherwise models are developed in a vacuum. We do not expect that work from experimental immunology will feed directly into the development of algorithms, but rather assist in the validation of immuno-engineering principles, which will then feedback into the development of models which then influences the library development. We now discuss possible avenues for experimental work in the context of immuno-engineering.

5.3.1 Cytokine Interactions

As a starting point for part experimental work, we would propose the testing of cytokine interaction models which can be done using in-vitro systems. In such experiments, control of the T-helper subset balance by innate signals derived from allergen can de demonstrated: proteolytically active allergen favoured the development of T-cells producing of IL4 (T-helper 2, Th2), and reduced numbers of interferon gamma producing T- cells (T-helper 1, Th1).

5.3.2 Regulatory Interactions

From insights gained from the modeling and immuno-engineering phases, these models can be developed to study regulatory interactions between cytokine producing subsets of cells. Multi-parameter flow cytometry and intracellular staining can be used (which allows visualisation of actual immune cells in a test-tube) to analyse cytokine profiles at the single cell level. Such techniques are able to analyse 6 colours of fluorescence simultaneously, allowing detailed characterisation of large populations (at least 10^6 individual cells) of responding lymphocytes. This approach has been used to model signalling molecule relationships [39]. The complexity of cytokine mediated control mechanisms has recently been emphasised by the description of a new helper T-cell subset producing interleukin 17 (Th17), which may have a critical role in immune mediated tissue damage and cancer [40]. Modelling of these complex interactions may provide a basis for intervention in this important group of diseases

5.4 Case Studies and a Comment on Applications of AIS

The case studies should be used to utilise and evaluate the architectures and algorithms developed as part of the above process. In combination, such case studies should display a range of immuno-informatics characteristics e.g. virtual, physical, open, dynamic and a range of space and time scales. The diversity of these characteristics should ensure that after case study development and feedback, the applicable scope of the immuno-engineering library will be sufficiently broad to support a wide range of applications.

Hart and Timmis [41] state that considering the application areas to date, AIS have been reasonably successful but, as yet, do not offer sufficient advantage over other paradigms available to the engineer. To address this and therefore tap the unexploited potential of AIS, one of the suggestions they make is that life-long learning is a key property of the immune system but true life-long learning, whereby a system is required to improve its performance as a consequence of its lifetime's experience, has not been utilised in AIS. Hart and Timmis propose a list of features they believe

AIS will be required to possess a combination of, if the field of AIS is to carve out a computational niche. These future AIS features, quoted verbatim from [21], are:

1. They will exhibit *homeostasis*
2. They will benefit from interactions between *innate* and *adaptive* immune models
3. They will consist of *multiple, interacting components*
4. Components can be easily and naturally *distributed*
5. They will be required to perform *life-long learning*

As for the future roles of AIS, Garrett [17] states that the biggest difficulty facing AIS is the lack of application areas to which it is clearly the most effective method. It is suggested that hybrid AIS may help to provide more powerful methods to solve certain problems. The current types of AIS used are also classified by Garrett into those that detect antigens (negative selection and danger theory models), and those that focus on destroying them (clonal selection and immune network models). It is pointed out, however, by Garrett [17] that the immune system has more to offer than this, with the mechanisms of the innate immune system and the view that the immune system is a homeostatic control system, being highlighted as future areas for AIS to exploit, and indeed this call seems to be taken up in a small part in recent times [42, 43, 44]. Concurring with the above, Bersini [45] argues that the immune system is much more than a simple classifier and performs much more than "pattern matching" and urges people in AIS to think about applications that are far removed from such applications which are the dominant force in AIS [23]. This challenges the community to find that *niche* application that AIS alone can tackle. This may come in the form of certain engineering type applications such as robotics and real-time systems where the system is embodied in the world and needs to be able to cope with extreme challenges that are constantly changing. Adopting an Immuno-engineering approach should enable us to begin to tackle this challenges.

6 Conclusions

In this paper, we have presented a new way of thinking about the development of immune-inspired systems and we have proposed the Immuno-engineering approach. Currently, like many bio-inspired paradigms, Artificial Immune Systems (AIS) are pale counterparts to their natural system. This is not to say that AIS should be like an immune system or copy exactly what the immune systems does, this would lead to not only complexity issues when instantiating such AIS, but also conceptual issues: just because the immune system does something it does not mean that we should do the same in an engineering context. Rather, we advocate the interdisciplinary interaction to develop biologically grounded, theoretically understood and well tested Immuno-engineering principles that can be deployed in a wide variety of application domains. Should such an Immuno-engineering approach be developed, we believe that AIS will then begin to not only capture the computationally interesting properties of the immune system, but be able to make a significant contribution to the immunology that serves as its inspiration.

Acknowledgements

The authors would like to thank Paul Andrews, Alex Freitas, Andy Greensted, Colin Johnson, Nick Owens, Jamie Tycross and Joanne White for input into some of the ideas behind Immuno-engineering. Jon Timmis would like to acknowledge the EP-SRC grant EP/E005187/1[2] during which some ideas in the paper have been developed.

References

1. Timmis, J., Andrews, P.S., Owens, N., Clark, E.: An interdisciplinary perpective on artificial immune systems. Evolutionary Intelligence **1**(1) (2008) 5–26
2. Cohen, I.R.: Real and artificial immune systems: Computing the state of the body. Imm. Rev. **7** (July 2007) 569–574
3. de Castro, L.N., Timmis, J.: Artificial Immune Systems: A New Computational Intelligence Approach. Springer (2002)
4. Forrest, S., Perelson, A., Allen, L., R.Cherukuri: Self-nonself discrimination in a computer. In: IEEE Symposium on Research in Security and Privacy, Los Alamos, CA, IEEE Computer Society Press (1994)
5. Ishida, Y.: Fully distributed diagnosis by pdp learning algorithm: Towards immune network pdp model. In: Proc. of the Int. Joint Conf. on Neural Networks. (1990) 777–782
6. Bersini, H., Varela, F.J.: Hints for adaptive problem solving gleaned from immune networks. In Schwefel, H., Manner, R., eds.: Proc. of the First Conference on Parallel Problem Solving from Nature. Springer-Verlag, Berlin, Germany (1991)
7. Timmis, J., Bentley, P.J., eds.: Proceedings of the 1st International Conference on Artificial Immune Systems (ICARIS 2002), University of Kent Printing Unit (2002)
8. Timmis, J., Bentley, P., Hart, E., eds.: Proceedings of the 2nd International Conference on Artificial Immune Systems (ICARIS 2003), LNCS 2787, Springer (2003)
9. Nicosia, G., Cutello, V., Bentley, P.J., Timmis, J., eds.: Proceedings of the 3rd International Conference on Artificial Immune Systems (ICARIS 2004). LNCS 3239, Springer (2004)
10. Jacob, C., Pilat, M., Bentley, P., Timmis, J., eds.: Proc. of the 4th International Conference on Artificial Immune Systems (ICARIS). Volume 3627 of Lecture Notes in Computer Science., Springer (2005)
11. Bersini, H., Carneiro, J., eds.: Proc. of 5th International Conference on Artificial Immune Systems. Lecture Notes in Computer Science, Springer (2006)
12. de Castro, L.N., Von Zuben, F.J., Knidel, H., eds.: Proceedings of the 6th International Conference on Artificial Immune Systems. Volume 4628 of Lecture Notes in Computer Science. Springer (2007)
13. Dasgupta, D., ed.: Artificial Immune Systems and their Applications. Springer (1999)
14. de Castro, L.N., Von Zuben, F.J.: Artificial immune systems: Part I—basic theory and applications. Technical Report DCA-RT 01/99, School of Computing and Electrical Engineering, State University of Campinas, Brazil (1999)

---

[2] http://www.bioinspired.com/research/xArcH/index.shtml

15. de Castro, L.N., Von Zuben, F.J.: Artificial immune systems: Part II—a survey of applications. Technical Report DCA-RT 02/00, School of Computing and Electrical Engineering, State University of Campinas, Brazil (2000)

16. Ji, Z., Dasgupta, D.: Artificial immune system (AIS) research in the last five years. In: Congress on Evolutionary Computation. Volume 1., Canberra, Australia, IEEE (December 8–12 2003) 123–130

17. Garrett, S.: How do we evaluate artificial immune systems? Evolutionary Computation **13**(2) (2005) 145–177

18. Timmis, J.: Artificial immune systems: Today and tomorow. Natural Computing **6**(1) (Feb. 2007) 1–18

19. Timmis, J., Knight, T.: Artificial immune systems: Using the immune system as inspiration for data mining. In: Data Mining: A Heuristic Approach. Idea Group (2001) 209–230

20. Kim, J., Bentley, P., Aickelin, U., Greensmith, J., Tedesco, G., Twycross, J.: Immune system approaches to intrusion detection - a review. Natural Computing **in print** (2007)

21. Hart, E., Timmis, J.: Application areas of AIS: The past, the present and the future. Applied Soft Computing **8**(1) (2008) 191–201 In Press, Corrected Proof, Available online 12 February 2007.

22. Timmis, J., Hone, A., Stibor, T., Clark, E.: Theoretical advances in artificial immune systems. Journal of Theoretical Computer Science **In press**(doi:10.1016/j.tcs.2008.02.011) (2008)

23. Forrest, S., Beauchemin, C.: Computer Immunology. Immunol. Rev. **216**(1) (2007) 176–197

24. de Castro, L.N., Von Zuben, F.J.: Learning and optimization using the clonal selection principle. IEEE Transactions on Evolutionary Computation **6**(3) (2002) 239–251

25. Gonzalez, F.A., Dasgupta, D.: Anomaly detection using real-valued negative selection. Genetic Programming and Evolvable Machines **4**(4) (2003) 383–403

26. Neal, M.: Meta-stable memory in an artificial immune network. [8] 168–180

27. Greensmith, J., Aickelin, U., Cayzer, S.: Introducing dendritic cells as a novel immune-inspired algorithm for anomaly detection. [10]

28. Aickelin, U., Bentley, P., Cayzer, S., Kim, J., McLeod, J.: Danger theory: The link between AIS and IDS? [8] 147–155

29. Bentley, P.J., Greensmith, J., Ujjin, S.: Two ways to grow tissue for Artificial Immune Systems. [10] 139–152

30. Twycross, J., Aickelin, U.: Towards a conceptual framework for innate immunity. [10] 112–125

31. Greensmith, J., Aickelin, U., Twycross, J.: Articulation and clarification of the dendritic cell algorithm. [46] 404–417

32. Orosz, M.: An Introduction to Immuno-Ecology and Immuno-Informatics. In: Design Principles from the Immune System. Sante Fe (2001) 125–150

33. Stepney, S., Smith, R., Timmis, J., Tyrrell, A., Neal, M., Hone, A.: Conceptual frameworks for artificial immune systems. Int. J. Unconventional Computing **1**(3) (2006) 315–338

34. Freitas, A., Timmis, J.: Revisiting the foundations of artificial immune systems for data mining. IEEE Trans. Evol. Comp. **11**(4) (2007) 521–540

35. Milner, R.: Communicating and Mobile Systems: the π-Calculus. Cambridge University Press (1999)

36. Phillips, A., Cardelli, L.: Efficient, correct simulation of biological processes in the stochastic pi-calculus. In: Proceedings of Computational Methods in Systems Biology (CMSB'07). Volume 4695. (2007) 184–199

37. Alon, U.: Uri alon, network motifs: theory and experimental approaches. Nature Reviews Genetics **8** (2007) 450–461

38. Gamma, E., Helm, R., Johnson, R., Vlissides, J.: Design Patterns. Addison Wesley (1995)

39. Sachs, K., Perez, O., Pe'er, D., Lauffenburger, D., Nolan, G.: Causal protein-signaling networks derived from multiparameter single-cell data. Science **308** (2005) 523–529

40. Steinman, L.: A brief history of t(h)17, the first major revision in the t(h)1/t(h)2 hypothesis of t cell-mediated tissue damage. Nature Medicine (2007) 139–145

41. Hart, E., Timmis, J.: Application areas of AIS: The past, the present and the future. [10] 483–497

42. Owens, N., Timmis, J., Greensted, A., Tyrrell, A.: On immune inspired homeostasis for electronic systems. [12] 216–227
43. Davoudani, D., Hart, E., Paechter, B.: An immune-inspired approach to speckled computing. [12] 288–299
44. Guzella, T., Mota-Santos, T., Caminhas, W.: Towards a novel immune inspired approach to temporal anomaly detection. [12] 119–130
45. Bersini, H.: Immune system modeling: The OO way. [46] 150–163
46. Bersini, H., Carneiro, J., eds.: Proceedings of the 5th International Conference on Artificial Immune Systems. Volume 4163 of LNCS. Springer (2006)

Heuristics for Uninformed Search Algorithms in Unstructured P2P Networks Inspired by Self-Organizing Social Insect Models

Prithviraj Dasgupta and Erik Antonson

Computer Science Department, University of Nebraska, Omaha, USA.
E-mail: pdasgupta@mail.unomaha.edu

Abstract We consider the problem of rapidly searching for resources or files in a distributed, unstructured, peer-to-peer file sharing network. Unstructured p2p network protocols such as Gnutella use a flooding-based mechanism for resource searching that generates considerable traffic in the network for each search query. When the searching activity by users in a p2p network is high, the traffic generated from the search requests could ensue congestion and result in increased search latency and poor performance in the entire network. To address this problem, we describe a resource search algorithm for p2p networks inspired by the stigmergetic behavior of ants while searching for food. Ants are used to encapsulate a search query initiated by a user in the p2p network. To search for the resource corresponding to their search query among the nodes of the network, each ant associates a certain amount of virtual pheromone with the nodes it visits. Later on, ants searching for resources use the amount and type of pheromone associated by previous ants with each node along their search path to direct the search query towards nodes that have a higher probability of resulting in the success for the search. We have tested our algorithm extensively within a simulated p2p network. Our simulation results show that our ant-based heuristics perform better than a completely uninformed or blind search that requires similar message overhead for each search query. When compared to a flooding-based mechanism, although the ant based search heuristic performs less efficiently under certain circumstances, it is capable of reducing the message overhead per search query by an exponential amount with respect to the flooding-based mechanism.

Keywords: Swarm intelligence, software agents, peer-to-peer networks, resource searching.

Please use the following format when citing this chapter:

Dasgupta, P. and Antonson, E., 2008, in IFIP International Federation for Information Processing, Volume 268; *Biologically-Inspired Collaborative Computing*; Mike Hinchey, Anastasia Pagnoni, Franz J. Rammig, Hartmut Schmeck; (Boston: Springer), pp. 19–32.

1. Introduction

Over the past few years, large scale distributed systems that can dynamically change their configuration over time and exhibit complex interactions between the system's components have emerged as an attractive paradigm for building robust, dynamic and adaptive systems. Such distributed systems have been used to design systems in diverse areas including autonomous multi-robot systems for unmanned search and rescue operations [1], peer-to-peer overlay networks for connecting millions of users for rapid interactions [2], data mining for bioinformatics applications [3], and pricing in online economies. Recently, several researchers have developed self emergent techniques inspired from disciplines such as biology to analyze the interactions between the components of such distributed systems and develop simple rules that control and dynamically adapt the system's behavior [4, 5]. In this paper, we consider the problem of rapidly searching for resources to reduce the latency facing users in a large scale, distributed, unstructured, peer-to-peer (p2p) network. A p2p network is an overlay network of nodes that allows users to share files and resources with one another. One of the major services employed by users in a p2p network is to search for and possibly download resources or files available with users on other nodes. With users and nodes in the order of millions in commercial p2p networks, rapidly searching for resources is a crucial problem that reduces the latency for all the users of the network as well as diminishes the traffic and congestion in the network.

Currently, in unstructured p2p networks, an uninformed search algorithm is used to locate resources being searched for by users across the different nodes of the p2p network. However, uninformed search is inefficient because it generates considerable traffic and congestion in the network through message flooding. In this paper, we describe a heuristics-based algorithm for searching for resources in an unstructured p2p network inspired by the stigmergetic behavior of social insects such as ants. In our algorithm, ants use virtual pheromone to direct a search query towards nodes that are likely to contain the resource being searched for in the search query. Each ant is implemented as a message that encapsulates the search query. An ant visits different nodes while searching for the resource encapsulated within its search query and associates a certain amount of virtual pheromone with the nodes visited by it. An ant searching for a resource later on uses the cumulative amount of pheromone left behind on nodes by previous ants to adapt its search behavior and direct its movement towards nodes that have a higher probability of containing the resource it is searching for. We employ different types of pheromone and different types of ants to improve the efficiency of the p2p search mechanism. Our simulation results show that our ant-based heuristics perform better than a completely uninformed or blind search that requires similar message overhead as our ant based algorithm for each search query. When compared to a flooding-based mechanism, the ant based heuristic evidently performs less efficiently but is able to achieve an exponential reduction in message overhead per search query as compared to the flooding mechanism.

2. P2P Resource Discovery Protocol

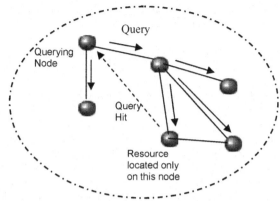

Figure 1 Messages exchanged in the p2p resource discovery protocol.

A p2p network consists of an interconnected collection of nodes. Nodes can join and leave the network dynamically. Each node contains resources that can be downloaded by other nodes in the network. Each node of a p2p network is usually associated with a human user that creates the node. One of the immensely popular applications of p2p networks, popularized by services such as Napster and Kazaa [6, 7, 8, 9], has been to enable file-sharing between the users of the p2p network. In a file sharing p2p network, the resources located on different nodes of the network are media files containing audio, video, image and even textual data. The major operation performed by a user in such a file sharing p2p network is to search for and download files available on other nodes in the network. Each node/user simultaneously allows other users/nodes to access the files present on the node itself. The searching for files (resources) in a p2p network is realized through the p2p resource discovery protocol shown in Figure 1. In the p2p resource discovery protocol, a user at a node wishing to search for a resource creates a search query containing certain unique keywords (e.g. filename or file tags) related to the resource it wishes to search for. The search query is encapsulated within a query message and provided with a search boundary that corresponds to the maximum number of hops the query message should be forwarded for, measured from its source. The *query* message is then forwarded to different nodes in the network in a breadth-first manner originating from the node that initiated the query. When a node in the p2p network receives the query message, a local search is performed among the resources present on that node. If the node contains a resource that returns a match with the keywords contained in the query message, the node sends a *queryHit* message to the node that originated the query, and does not forward the query message to other nodes. On the other hand, if none of the resources on the node that receives the query message returns a match for the keywords in the query message, the query message is forwarded to all the neighbors of that node, provided the search boundary of the query message has not been reached. After a certain period of time since sending out the search query, the user at the node originating the search query observes all the queryHit messages it has received from other nodes in the p2p network and selects one or more of these nodes to download the resource from.

A potential source of inefficiency in the p2p resource discovery protocol is the enormous amount of messages generated in a p2p network for each search query. For example, if the average degree of a node in a p2p network is denoted by 'd' and the search boundary of a query message is denoted by 'b', each query message results in d^b query messages in the p2p network, in the worst case. With several million users simultaneously using a p2p network, the overall traffic generated from query messages in a p2p network can be overwhelming. This adversely affects the performance in the network by generating enormous amounts of network traffic ensuing congestion. In this paper, we posit that the traffic from query messages in a p2p network can be reduced if the breadth-first traversal used by the query messages in the resource discovery protocol is replaced by a local heuristics-based search that generates less than the exponential number of query messages per search query. Clearly, if we reduce the number of messages generated by a search query, the search query reaches fewer nodes and could potentially result in lower success for the search. Therefore, the heuristics must be carefully designed to compensate for the potential loss in reachability to nodes with an intelligent node selection strategy that forwards to query message to a few neighbor nodes that improve the probability of finding the resource on or along them. To design such a heuristic, we have used the stigmergetic behavior of social insects such as ants in locating objects of interest such as food.

2.1   Ant- Algorithm Based Resource Discovery in P2P Networks

Stigmergy is a process that enables insects to communicate information with each other either directly (e.g., by physical contact) or indirectly (e.g., by depositing chemical trails in the environment)[4]. The communication of information through stigmergy results in a swarm-like collective, emergent behavior between the insects to achieve complex tasks in a collaborative manner. For example, ants searching for food initially explore the environment around their nest. Each ant leaves behind a trail of a chemical substance called pheromone. Pheromone serves as an attractor for ants searching for the food later on. Pheromone also evaporates with time, to model the volatility in the environment and enable ants search for different locations for food as well as to enable the gradual removal of trails after the food at a location is exhausted. When an ant locates food at a particular location, it returns back to the nest while depositing pheromone on the ground. Consequently, the trail that leads from the nest to the food receives the maximum amount of pheromone and ants get attracted towards it. These ants further reinforce the pheromone trail between the nest and the food and enable ants to reach the food from the nest thereafter by following the pheromone trail.

To design the heuristics for the p2p resource discovery protocol based on the stigmergetic behavior of ants, we have modeled the forwarding of a query message as the movement of a virtual ant between the nodes visited by the query message. The stigmergetic behavior of these virtual ants is realized through virtual pheromone associated with the nodes that each ant visits. To record the pheromone information at each node, we maintain a pheromone table within each node. The pheromone table of a node contains the address or identifier of each neighbor of that node and a real value corresponding to the amount of pheromone associated with that neighbor node. Initially, the pheromone value of each neighbor node is initialized to zero. As ants search for resources, they update the values in the pheromone table of each node they visit while searching for resources. When an ant reaches a node, it determines the neighbor node to visit next based on the pheromone value inside the pheromone table of the node.

In a p2p network, nodes can leave and join the network in an ad-hoc manner. Also, a user at a node can dynamically add and remove resources that are stored within the node. This dynamic nature of a p2p network and the resources within its nodes implies that the pheromone values within the pheromone tables of the different nodes must be dynamically updated to enable ants visit newly joined nodes to search for resources within them, as well as to remove outdated trails that lead to removed resources or nodes that have already left the network. Consequently, the ant algorithm needs to balance the exploitation of existing trails to search for resources with exploration of nodes to discover newly added resources. To achieve this balance between exploitation and exploration in our algorithm, we have used two different types of pheromone in our model. The first type of pheromone, called *pheromone*, is used to mark routes that have resulted in successful searches and enables exploitation of existing trails. The second type of pheromone, called *anti-pheromone*, is used to mark nodes that have resulted in an unsuccessful search and enables exploration of new nodes or nodes that did not have resources that were searched for in the past. Based on these two types of pheromone, we have also designed two different types of forward-moving ants for our algorithm, which exhibit different responses to the different types of pheromone. The different ant types used in our algorithm are described below:

- **Forward Foraging Ants.** These ants deposit pheromone at nodes they visit and also get attracted to higher amounts of pheromone and repelled by higher amounts of anti-pheromone. From each node, a forward foraging ant prefers to go to neighbor nodes that have higher amounts of pheromone and lesser amounts of anti-pheromone.
- **Forward Explorer Ants.** These ants deposit anti-pheromone at nodes they visit and also get attracted to higher amounts of anti-pheromone and repelled by higher amounts of pheromone. From each node, a forward explorer ant prefers to go to neighbor nodes that have higher amounts of anti-pheromone and lower amounts of pheromone.

- **Backward Ants** Both types of forward ants become a backward ant when either they discover the resource on a node, or, they reach the search boundary without discovering the resource. Backward ants trace the route taken by their corresponding forward ant in the reverse direction. A backward ant deposits pheromone at each node it visits along its route, if the resource that the corresponding forward ant was looking for was found, and, deposits anti-pheromone at each node it visits along its route, if the resource was not found.

A forward ant contains the search algorithm that is executed by the ant on arriving at a node. The search algorithm performs a linear search for the terms in the user's search query encapsulated by the ant over the resources present on a node and returns a success only if there is a match.

3. Model

Our model of the p2p network comprises a connected network of N nodes. Each node contains certain resources inside a resource table and a resource can be identified on a node with a unique identifier. Nodes join and leave the network at random. Each node maintains a forwarding table containing the addresses of its neighbor nodes determined using the p2p node discovery protocol. Each address in the forwarding table is associated with a normalized weight that represents the probability of ant to migrate to that node. The weight of a node in the forwarding table gets updated when an ant selects it to move to it. Pheromone increases the weight while anti-pheromone decreases it. The use of a single weight attribute to reflect both types of pheromone keeps the ant algorithm simple and also reduces the size of the forwarding table.

A user at a node initiates a search by providing a set of keywords corresponding to the identifier of the resource(s) he or she wishes to locate. The node originating the query creates a forward ant with an empty stack. At each node, the ant selects a neighboring node to move to for the next hop, with a probability given by the weight of the node in the forwarding table. Before migrating to the selected node, the ant updates the weight of the node in the forwarding table according to the ant's type as described below.

Table 1 Parameters used for the ant-heuristics based p2p resource discovery alogrithm

Symbol	Parameter
a_n	Number of nodes in the forwarding table of node n
$w_{i,n}^{t}$	Normalized weight associated with neighbor node i of node n at time t
τ_n	Amount of pheromone deposited on node n
τ_0	Amount of pheromone deposited by an ant at the source node of the search
χ_n	Amount of anti-pheromone deposited on node n
χ_0	Amount of anti-pheromone deposited by an ant at the node on which search terminated
$h_{s,n}$	Number of hops made by an ant to reach from the node s on which it started its journey to the current node node n

Forward Foraging Ant. A forward foraging ant starts from its origin with an empty internal stack. The algorithm used by a forward foraging ant at a node n to select a neighbor node i and update the weight associated with node i uses the following parameters shown in Table 1. The update rules for the pheromone at node n are the following:

$$\tau_n = \frac{\tau_0}{[h_{s,n}]^\alpha}$$
$$w_{i,n}^t = w_{i,n}^{t-1} + \tau_n(1 - w_{i,n}^{t-1}) \tag{1}$$

and,

$$w_{i,n}^t = \frac{w_{i,n}^t}{\sum_{i=1}^{i=a_n} w_{i,n}^t} \tag{2}$$

The factor α is determined experimentally and it controls the decrease in the amount of pheromone deposited as the ant moves further away from its origin. The second term on the r.h.s of Equation 1 ensures that the amount of pheromone deposited on a node is proportional to its current weight. This prevents excessive pheromone (or anti-pheromone) being deposited on a node whose weight is very high (or low). Equation 2 ensures that the weights of nodes in the forwarding table remain normalized after the weight of a node is updated by an ant. The ant pushes the address of the current node into its internal stack before moving to the selected node.

Forward Explorer Ant. A forward explorer ant works in a manner similar to a forward foraging ant except the following:
- It uses the inverse probability $(1 - w_{i,n}^t)$ to select a node i from the forwarding table of its current node n. This ensures that the probability of

selection of a node by an explorer ant is proportional to the amount of anti-pheromone deposited on it.

- It updates the anti-pheromone at each visited node according to the following equations:

$$\chi_n = \frac{\chi_0}{[h_{s,n}]^\alpha}$$

$$w_{i,n}^t = w_{i,n}^{t-1} - \chi_n(1 - w_{i,n}^{t-1}) \tag{3}$$

where s is the origin node for the explorer ant.

Backward Ant. When the forward ant locates a resource or reaches its search boundary without locating the resource, it becomes a backward ant. The backward ant inherits the stack from its corresponding forward ant. If the resource was located by the forward ant, the backward ant rewards each node along the reverse route with pheromone using Equation 2. Otherwise, if the search boundary was reached without locating the resource, the backward ant deposits anti-pheromone on each node it visits using Equation 3 to indicate that the node did not lead to a successful resource discovery. For the backward ant, the node s represents the node on which the resource was found (in Equation 2) or the node on which the search boundary was reached without locating the resource (in Equation 3).

4. Simulation Results

We have implemented a Java application to simulate a dynamic p2p network and verify the performance of our ant-based p2p resource discovery algorithms. The Java application implements each node in the p2p network as a thread that is capable of communicating with each other threads (nodes) via message passing. To simulate the dynamic joining and leaving of nodes in the p2p network, we have used the 'churn' parameter that controls the rate at which nodes enter and leave the p2p network. Also, to simulate the availability of resources across the nodes of the p2p network, we have assumed that the probability of locating a resource at a particular node by an ant searching for the resource is a function of the number of resources available at that node. The number of resources available at a particular node is determined using a zipf distribution [10]. The default values of the different parameters used in our simulations are shown in Table 2. In the simulation experiments, we vary different parameters of our ant-based algorithm and in the p2p network and compare the effect of varying the parameters on the success ratio for the search, and, compare the performance of the ant-based algorithm with the two comparison strategies. For all our results, we have used the ratio between the number of successful search queries and the total number of search queries (called success ratio) as a measure of the performance of the search algorithm. All results are averaged over 10 simulation runs.

We have compared the performance of our ant-based algorithm for resource searching with two other search techniques. First, we have used a random or uninformed search mechanism where, at every hop, a search query is forwarded to one of the neighbors of the current node selected at random (denoted by legend R in the graphs of the simulation results). Since a search query is forwarded to only one node at every hop in this random mechanism and our ant-based mechanism, both these mechanisms generates at most d messages for each search query, where d is the search boundary for the search query. For our second comparison strategy, we have used the breadth first search (BFS) strategy currently used for resource searching in unstructured p2p network protocols such as Gnutella (denoted by legend BFS in the graphs of the simulation results). The BFS strategy forwards the search query to all neighbors of a node at every hop. Therefore, the message overhead of the BFS strategy for a single search query is b^d where b is the average node degree in the p2p network and d is search boundary for the search query.

Table 2 P2P Network Parameters for the Simulations

Name	Parameter	Value
N	Number of nodes in the p2p network	{200, 500, 1000}
D	Number of neighbors per node (average node degree)	4
Churn	Rate at which nodes join and leaving the network	50
ρ	Resource availability (Probability of a search ending successfully at a node)	0.02
numberOfAnts	Number of ants or search queries originated by each node during the lifetime of the simulation	numberOfFiles * ρ / 4 = 25
B	Search boundary for each ant	4
P	Probability of foraging	{0.0, 0.2, 0.4, 0.6, 0.8, 1.0}
χ	Amount of anti-pheromone deposited on the origin node/search boundary	0.6
τ	Amount of pheromone deposited on the origin node/resource node	0.6
α_f	Decrease in the amount of pheromone deposited as the ant moves away from the origin node	4.0
α_b	Decrease in the amount of pheromone deposited as the ant moves away from the search boundary/resource node	4.0

4.1.1.   Varying the Weight of Pheromone

For our first simulation, we observe the effect of varying the amount of pheromone deposited by forward foraging ants on the different nodes visited along their search path (τ) between 0.3 to 0.9. As shown in Figure 2(a), we observe that varying the pheromone weight has a limited effect of about 0.2% in the success ratio of search queries. The largest amount of pheromone, $\tau = 0.9$, is the most effective overall and this can be attributed to the fact that larger amounts of pheromone require a longer time to decay and are able to sustain successful trails over a longer period of time , thereby causing subsequent ants to follow those successful trails and locating the resources they are searching for. A similar result was obtained on the success ratio by varying the amount of anti-pheromone deposited by explorer ants on nodes along their search path.

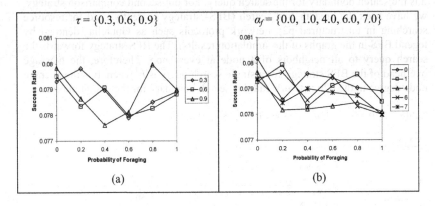

Figure 2 (a) Success Ratios for Varying Pheromone Weights. (b) Success Ratios for Varying Forward Ant Pheromone Decay Rates.

4.1.2.   Varying the Spatial Decay of Pheromone

For our next set of simulations, we varied the decrease in the amount of pheromone deposited by each forward ant as the ant moves away from the origin node (α_f) between 0.0 and 7.0. As shown in Figure 2(b), we observe that varying the forward ant pheromone decay rate has a limited effect of about 0.2% on the success ratio of the search queries. The smallest pheromone decay rate, $\alpha_f = 0.0$ (no decay), is the most effective overall. This behavior can be attributed to the fact that larger values of pheromone decay rate result in pheromone trails decaying rapidly at nodes further away from a successful node. Consequently, with a higher pheromone decay rate ants are not able to locate successful nodes and direct their search effectively when they are even a few hops away from a successful node. A similar result was obtained for the anti-pheromone decay rate (α_b).

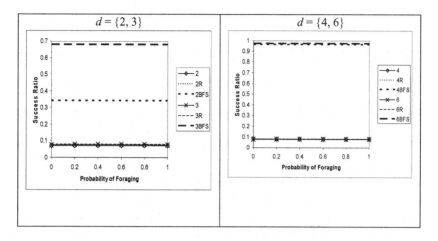

Figure 3 Success Ratios for Varying Neighbors with 200 nodes (left graph) and 1000 nodes (right graph).

4.1.3. Varying the Number of Nodes and the Number of Neighbors

For the simulations in this group, we varied the number of neighbors per node (d) between two or three (left graph) for a network with 200 nodes, and, between four and six (right graph) for a p2p network with 1000 nodes. As shown in Figure 3, we observe that d = 3 in a network of 200 nodes results in a higher success ratio because the network is more connected. However, in a network of 1,000 nodes changing the degree of each node from 4 to 6 has very limited effect on the success ratio. The reason for this can be attributed to the fact that increasing the node degree in the network beyond an average node degree of 4 does not significantly affect the network diameter and consequently does not improve the success ratio for the search. In fact, for our ant-based algorithm, increasing the node degree beyond a certain value obfuscates efficient trails and reduce the success ratio.

4.1.4. Varying the Resource Availability

For the next set of simulations, we vary the resource availability (ρ) on a node between 0.005 and 0.1. However, all nodes in the network have the same resource availability. As shown in Figure 4(a), we observe that varying the resource availability has a significant effect of about 36% on the success ratio for the search queries. The largest resource availability, $\rho = 0.1$, is evidently most effective because increasing the number of resources in the network reduces the search time required to locate a resource. In the next set of simulations, we verify the effect of heterogeneous values of the resource availability parameter. ρ is set to different values on different nodes of the network following the values of resource availabilities in actual p2p networks reported in [10]. Here, ρ varies between

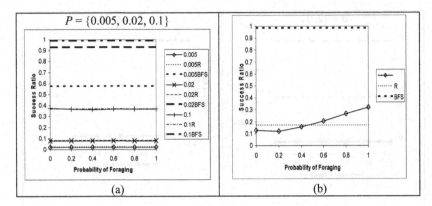

(a)   (b)

Figure 4 (a) Success Ratios for Varying Resource Availabilities (b) Success Ratios
for Heterogeneous Resource Availability.

0.0005 (on 30% of nodes), to 0.0055 (on 30% of nodes), to 0.055 (on 30% of
nodes), and 0.2525 (on 10% of nodes). As shown in Figure 4(b), we observe that
varying the probability of foraging for this p2p network configuration has a
significant effect of about 21% in the success ratio of the search queries. The ant-
based search behavior was more successful than the random search behavior for
probabilities of foraging of 60% and greater.   This would suggest that p2p
networks having a high degree of heterogeneity will benefit the most from ant-
based heuristics for resource discovery within the network. Moreover, it should
be noted that a probability of foraging of 100% appears to be the most effective
for this type of network.

For the next set of simulations, the resource availability (ρ) on different nodes is
allowed to vary between 0.01 and 0.7. The p2p network for this simulation has
1,000 nodes with six neighbors per node. As shown in Figure 5, we observe that
varying the resource availabilities for these p2p network configurations has a
significant effect of about 93% in the success ratio for the search queries. The ant-
based search behavior is more successful than the random search behavior for all
the distributions. The ant-based search makes its greatest improvement (up to a
15% improvement over the random search) at aggregate resource availabilities of
4% and 12%. The breadth-first search was significantly more effective than the
ant-based search for the heterogeneous resource distributions having fewer
resources; but, as the aggregate resource availability for the network simulations
increases, the success ratios for the ant-based search approaches the success ratios
for the breadth-first search.

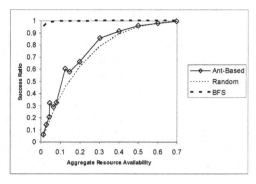

Figure 5 Success Ratios for Various Heterogeneous Resource Availabilities.

5. Related Work

Resource discovery is implemented by flooding a resource query across nodes of the network in most commercial p2p systems [7]. However, query flooding produces considerable network traffic by blindly forwarding the query across the network. Improvements to query flooding include p2p stacks [6, 11] that use super-peer nodes, and, dynamic hash table (DHT) based techniques that strategically place resources on nodes using a hash function to improve resource availability and enable rapid lookup [12, 13]. In contrast to our ant-based algorithms, these techniques focus more on resource management and do not incorporate the information obtained from previous resource queries to improve future searches. Ant algorithms have already been applied to several applications [3, 14] including dynamic programming, traveling salesman problem and routing in telecommunication networks [5]. However, resource discovery in p2p networks is different from each of these applications because the node on which the resource will be discovered is not known *a priori* and the topology of the p2p network can change dynamically as nodes join and leave. Extensions to ant algorithms using anti-pheromone for the traveling salesman problem have been studied in [15]. The Anthill framework[2] employs ant-based algorithms for load balancing in a p2p network and ants backtrack along the path they traveled to update routing tables at each node. In contrast our algorithm uses different types of pheromone and ants with different behavior to make p2p resource discovery more efficient.

6. Conclusions and Future Work

In this paper, we have described an informed search algorithm using an ant-based heuristic for p2p resource discovery. We are currently investigating extensions to the algorithm described in this paper using multiple ants to enable parallel search queries. We are also exploring techniques that enable ants to dynamically change their type and the pheromone deposited by them based on the performance of their search query. Finally, we plan to develop techniques that allow ants to exchange trail information with each other to locate resources rapidly. We envisage that biology inspired emergent algorithms provide a useful direction for further exploring challenges and issues of p2p networks for future research.

References

1. H. Van Dyke Parunak, and S Brueckner : Swarming Coordination of Multiple UAVs for Collaborative Sensing. In: Proc. 2nd AIAA 'Unmanned Unlimited' Systems Conference, San Diego, CA, (2003).
2. O. Babaoglu, H. Meling and A. Montresor: Anthill:A framework for the development of agent-based peer-to-peer systems. In: Proc. 22nd International Conference on Distributed Computing Systems (ICDCS), pp. 15-22, (2002).
3. Abraham, C. Grosan, V. Ramos (eds.): Swarm Intelligence in Data Mining,"Studies in Computational Intelligence , vol. 34, Springer, (2006).
4. E. Bonabeau. , M. Dorigo, G. Theraulaz: Swarm Intelligence: From Natural to Artificial Systems. Oxford University Press, 1999.
5. G. Di Caro and M. Dorigo: AntNet: Distributed Stigmergetic Control for Communications Networks. Journal of Artificial Intelligence Research, vol. 9, pp. 317-365, (1998).
6. Fast Track, URL http://www.fasttrack.com
7. Gnutella, URL http://www.gnutella.com
8. Kazaa, URL http://www.kazaa.com
9. Napster Inc., URL http://www.napster.com
10. S. Saroiu, P. Gummadi, S. Gribble: Measuring and analyzing the characteristics of Napster and Gnutella hosts. Multimedia Systems, vol. 9 no.2, pp 170-184, (2003).
11. B. Yang and H. Garcia-Molina: Designing a super-peer network. Proc. 19th International Conference on Data Engineering (ICDE), pp. 49-62, (2003).
12. I. Stoica, R. Morris, D. Karger, F. Kaashoek, and H. Balakrishnan: Chord: A peer-to-peer lookup service for internet applications. In: Proc. ACM SIGCOMM Conference, pp. 149-160, (2001).
13. J. Kubiatowicz, et al.: OceanStore: An Architecture for Global-Scale Persistent Storage. In: Proc. ACM ASPLOS, pp. 190-201, (2000).
14. P. Dasgupta: Improving Peer-to-Peer Resource Discovery Using Mobile Agent Based Referrals. In: Proc. 2nd Workshop on Agent Enabled P2P Computing, pp. 41-54, (2003).
15. J. Montgomery, and M. Randall: Anti-pheromone as a tool for better exploration of search space. In: Lecture Notes in Computer Science, vol. 2463, Springer-Verlag, pp. 100-110, (2002).

Congestion Control in Ant Like Moving Agent Systems

Alexander Scheidler, Daniel Merkle, and Martin Middendorf

Abstract In this paper we study the problem of congestion in system where agents move according to simple ant inspired movement rules. It is assumed that the agents have to visit a service station to refill their energy storage. After visiting the service station the ants can move randomly and fast. The less energy an agent has the slower it becomes and the more it moves in direction of the service station. Different methods for self-organized congestion control are proposed in this paper where the behavior of the agents compared to the original is not changed or is changed only slightly without the need to use any global information and without using additional sensory information. The proposed systems are investigated with

1 Introduction

The (movement) behavior of social insects is an inspiring source of ideas for the design of methods for solving various problems in computer science and related fields. Examples are the well-known Ant Colony Optimization method as introduced in [1] (for an overview see [2]), algorithms for the movement behavior of robots [3], ant inspired clustering methods (for an overview see [4]), or the self-organized behavior of the compounds in Organic Computing (OC) [5, 6] systems (for an overview see [7]). Emergent pattern

Alexander Scheidler and Martin Middendorf
Department of Computer Science
University of Leipzig, Postfach 100920, 04009 Leipzig, Germany
{scheidler,middendorf}@informatik.uni-leipzig.de

Daniel Merkle
Departement of Mathematics & Computer Science
University of Southern Denmark, Campusvej 55, DK-5230 Odense M, Denmark
merkle@imada.sdu.dk

Please use the following format when citing this chapter:

Scheidler, A., Merkle, D. and Middendorf, M., 2008, in IFIP International Federation for Information Processing, Volume 268; *Biologically-Inspired Collaborative Computing*; Mike Hinchey, Anastasia Pagnoni, Franz J. Rammig, Hartmut Schmeck; (Boston: Springer), pp. 33–43.

that might arise when groups of animals move have been deeply investigated in biology (e.g., [8, 9, 10, 11]). In [12] the so called sorting behavior in the brood chambers of the ant *Leptothorax unifasciatus* has been investigated. In the brood chambers the youngest brood items (eggs and microlarvae) are placed in the chambers center, larger larvae are arranged in concentric rings around the center, and the largest and oldest brood (pupae and prepupae) is placed in an intermediate area between the peripheral larvae and the larvae of medium size. One explanation why this sorting occurred is that the brood distribution pattern helps to organize the brood care in the nest. It has been shown by simulation studies that a system with very simple behavioral rules for the movement of agents can show an emergent sorting behavior. The sorting behavior of ants has inspired the design of methods that are used by robots to solve sorting problems [13].

In [14] movement models that are inspired by the ant *Leptothorax unifasciatus* ants have been applied to OC systems with moving agents. In the studied OC systems the agents have one (or several) service stations which they have to visit from time to time (e.g., to recharge their batteries or to drop items they have collected). It was shown that emergent patterns that occur within the distribution of the agents with respect to the different service stations can occur even when only very slight behavioral differences between the agents exist. It was also shown that a problem with the ant inspired movement models is that unwanted congestion can emerge at the service stations unless there is only a small number of agents in the system.

In this paper we propose and study some methods to reduce and control the emergent congestion in ant like moving agent systems. The aim is to develop methods where the behavior of the agents is not changed or is changed only slightly without the need to use any global information and without using additional sensory information. While reducing congestion (and thereby increasing the performance of the system) it is important that the fairness of the system is not reduced. Fairness is measured here as the variance of the waiting times of the agents before they can visit the service station.

The paper is structured as follows. In the next Section 2 the agent model and the movement behaviour are introduced. The methods for congestion control are introduced in Section 3. The experiments are described in Section 4. Results are presented in Section 5. Conclusions are given in Section 6

2 Agent Model

Different models for the movement of ants within a nest have been introduced in [12]. It was shown that small differences in the movement behavior of the ants can lead to spatial sorting of the ants (i.e., on average over time ants with different behaviour can be found in different areas of the nest). The degree of the sorting depends on the particular movement model. In principle each

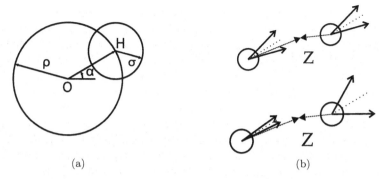

Fig. 1 (a) Every agent is modeled as a disc; ρ - radius; O - centre of the body (x, y); α - direction of movement; H - centre of the head; σ - sensing range. (b) Effect of different values of the parameter μ_i on the turning behavior when unobstructed; Z is the service point; (upper) for large μ_i there is only a slight difference between moving from or to the service point; (bottom) for small μ_i the turning angle becomes significantly smaller the smaller the angle between actual moving direction and the vector to the service point is

of the defined movement models consists of the following two actions for an ant: "turning" the movement direction and "moving" straight forward.

In [14] some of the movement models of [12] have been changed slightly to avoid an unnatural blocking effect that was observed and also to adapt the models to fit the requirements of OC systems. It was shown that emergent spatial sorting patterns for groups of randomly moving ant like agents can depend on slight differences in the movement models. It was observed that the movement of ants can depend on the CO_2 gradient which typically points to the center of a brood chamber ([15, 16]). It was argued in [14] that the movement behaviour of agents in OC systems might similarly depend on their movement direction with respect to the direction of a service station. But, different from the natural systems in the OC system there might be several service points that influence the movement of the agents. It was shown that the relative size of the influence area of the service points can lead to an interesting and strong emergent pattern within the spatial distribution of agents with slightly different moving behavior.

The movement models of the agents that are used in this paper correspond to the *repulsive ant model* from [14]. This model which is a mixture of the *centripedal ant model* and the *avoiding ant model* from [12]. This mixed movement model was introduced to overcome the problem that agents get stuck near the focal point (see [14] for details). The behavioural differences between the agents were modeled in the way that each agent i has a parameter $0 \leq \mu_i \leq 1$ that influences its moving behaviour. Fixed values $\mu_i = i/(n+1)$ for agent $i \in 1, \ldots, n$ were used for the experiments in [12] and [14]. In this paper we investigate a movement model where the parameter μ_i can vary over time for every agent. The movement behaviour is described in the following.

Shape of the Agents. Similar as in [12] the shape of an agent is modeled as a disc with radius ρ. The centre of the disc (x_i, y_i) represents the position of agent i. Each agent has an actual direction of movement α_i, which is measured as the angle relative to the lower border of the simulation area. The point at position $(x_i + \rho \cos \alpha_i, y_i + \rho \sin \alpha_i)$ models the centre of the agents head. From the centre of the head every agent can sense obstacles within a range of distance σ, called sensing range (see Figure 1). Agent i collides with agent j if agent j is within the sensing range of agent i, i.e., when the distance between the centre of the head of agent i and the centre of the body of agent j is smaller than $\sigma + \rho$. Similarly, an agent collides with the nest wall when the euclidian distance between the centre of its head and the wall is less than the sensing distance.

Movement when unobstructed. If there is no obstacle (wall or other agent) within its sensing range an agent will move and turn at each time step. The agent moves distance ν_i in direction α_i, i.e., $x_i \leftarrow x_i + \nu_i \cos \alpha_i$ and $y_i \leftarrow y_i + \nu_i \sin \alpha_i$. The different values ν_i represent different velocities of the agents. The parameter value ν_i dependents on the internal parameter μ_i of the agents as follows: $\nu_i = (1 - \mu_i)\nu_s + \mu_i \nu_f$ where parameters ν_s and ν_f, $0 < \nu_s < \nu_f < 1$ denote the slowest and the fastest velocity. Within the same time step the agent also makes a turn by changing its movement direction by $\alpha_i = \alpha_i + \theta_i$, where θ_i depends on the internal parameter μ_i of the agent and ϕ_i is the angle between the actual moving direction α_i and the vector towards the service point. For the calculation the clinotaxis model from [17] is used: $\theta_i \leftarrow p_u(1 - \mu_i)\chi + p_b \mu_i \tau \cdot (1 - \cos(\phi_i))/2$ where $\chi = 15°$, $\tau = 30°$. The values of p_u and p_b are randomly chosen from $\{-1, 1\}$ and determine the direction of turning. The turning behavior depends on ϕ_i and the larger this angle is the stronger the agent will turn. Agents with larger value μ_i will be less affected by their ϕ_i as agent with small μ_i (see Fig. 1 b). Therefore, for agents with small value μ_i the attraction to the service point is stronger than for agents with large value μ_i.

Movement when obstructed. If the wall or another agent is within the sensing range of an agent, the agent will not move, but only make a turn. It avoids the obstacle explicitly by turning into one direction until it can move again. To define the turning direction assume that agent i collides with agent j. The sign of the scalar product between the vector that is perpendicular to the vector of the moving direction of agent i and the vector from the centre of agent i to the centre of agent j determines the direction of turning: $\theta_i \leftarrow \text{sign}((-\sin \alpha_i, \cos \alpha_i) \cdot (x_j - x_i, y_j - y_i))U(0, \Theta_i)$. A collision with the nest wall is handled analogously.

In our model we use the parameter μ_i to model the state of an agent. The higher the value of μ_i is the faster can the agent move. Therefore, if an agent has visited the service area the value of μ_i is increased. The motivation behind this is that in applications the agents might get new power at the service station or has been unloaded at the service station. During movement of an agent its value μ_i decreases. The motivation is that in applications the agent might use power or pick up items.

3 Congestion Control

To resolve possible congestion of the agents at the service point we introduce three different congestion control methods. The goal of these methods is to resolve the congestion either by leaving the behaviour of the agents unchanged or by changing the behaviour of the agents only slightly but without need for introducing any new type of sensory information or global knowledge. The first two control methods C_P and C_W do not change the agent behaviour and the third method C_D changes only the sensing range of the agents.

The idea of control method C_P is to introduce two parallel walls next to the service station that form a pipe. The idea of this method is that agents that have visited the service station and have a hight value μ_i might be able to move away from the service station through the pipe whereas only few of the agents that have a small value μ_i might use the pipe to move to the service station. An example of a pipe can be seen in the left of Figure 3.

Control method C_W is to introduce two additional walls on two sides of the service station. Each wall has an small opening in the middle that is next to the service station. The idea of this method is that slow agents with small value of μ_i might be forced to wait behind a wall and therefore do not block the service station. Hence, the agents that have visited the service station can move away from it. An example for this control method C_W can be seen in the middle of Figure 3.

The third control method C_D is to change the behavior of the agents slightly. Here the sensing range σ_i of agent i depends on the internal parameter μ_i. The sensing range is calculated as follows: $\sigma_i = 2\rho - 1.4\mu_i\rho$. The idea behind this method is that agents with a small value of μ_i that move to the service station have a larger sensing range and therefore leave some space between them when they come next to the service station. This space can be used by the agents that have visited the service station and therefore have a large value μ_i can to move away from the service station.

4 Experiments

The simulation area is a quadratic field with side length 1. At the start of a simulation run the positions of the agents are distributed randomly with uniform distribution over this area. Also the values of the internal parameters μ_i are chosen randomly with uniform distribution between 0 and 1. In the centre of the field there is a circular service area with radius 0.04. The centre of this area is the service point. If an agents position (i.e., the centre of its body) is within the service area its internal parameter μ_i is set to 1. If agent i moves (e.g. the agent is unobstructed) the value of μ_i is decreased by a fixed value 0.001 until $\mu_i = 0$. Observe that the smaller the value of μ_i is the slower moves the agent and also the higher is the attraction force to the service point.

For the experiments the body of an agent has radius $\rho = 0.01$, the radius of the head is $\sigma = 0.006$. Parameters ν_s and ν_f that denote the slowest and the fastest velocity are set $\nu_s = 0.0006$ and $\nu_f = 0.006$. Per time step parameter μ_i of agent i reduced by 0.001. Different numbers of agents have been used and for each number of agents and each congestion control method each run was repeated 20 times over 10000 time steps each.

5 Results

Figure 2 shows the distribution of the agents after 2000 time steps for different number of agents. It can be seen that a system with 90 agents works without strong congestion at the service station. It can also be seen that agents with small value μ_i (bright color) tend to be close to the service station. Agents with large value μ_i are nearly randomly distributed over the whole field. This is different for a system with 150 agents. Here most nearly all agents can be found close to the service station. The agents with large value μ_i can be found or very near to the service station. They cannot move away because the way is blocked by the agents with small value μ_i that try to move into the service area. As shown later, for this system the agents cannot do much useful work (if that means that the agents should ideally move over the whole field). Altogether, the observed congestion is an unwanted effect of the system that depends on colony size.

Figure 3 shows the distribution of the agents for a system with 150 agents after 2000 time steps using one of the congestion methods C_P and method C_W. It can be seen that there is much less congestion by slow agents with small value μ_i within the pipe for method C_P than outside of the pipe next to the service area. It can also be seen that the agents with high value μ_i can move through the pipe.

For method C_W it can be seen that the congestion around the service area is much less compared to the system without congestion control. Agents with high value μ_i can be found in different parts of the filed and not only next to the service area (as it was the case when no congestion control is used.

For method C_D the distribution of the 150 agents after 2000 time steps is show in Figure 4. The figure shows that at least some agents with high value μ_i that have visited the service station can move away from it because the agents with small value μ_i leave some space between each other.

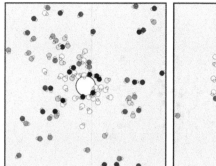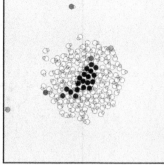

Fig. 2 Distribution of the agents for different number of agents after 2000 time steps; (left) 90 agents; (right) 150 agents; the smaller the value μ_i of an agent the brighter is its color; the service area is the white circle in the middle

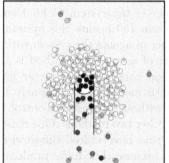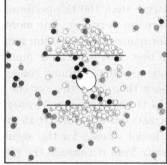

Fig. 3 Distribution of agents for a system with 150 agents after 2000 time steps using congestion method C_P (left), method C_W (right)

To compare the performances of the control methods with system that uses no control method the following measure for the performance of the system is used. If an agent reaches the service area its value μ_i is increased by adding the value $1 - \mu_i$ so that $\mu_i = 1$ holds afterwards. Summing up over all values $1 - \mu_i$ for all i and every time when the value μ_i is increased can be seen as measure for the performance the system. This value is called the total energy consumption of the system and is denoted P_T when measured over the first T time steps. Since agents that move use energy whereas agents that can not move do not use energy the total energy consumption is a measure how freely the agents can move on average.

Figure 5 shows the total energy consumption P_T for a system without congestion control and systems with congestion control. It can be seen that for a small number of agents when no congestion occurs the system without congestion control has the highest performance. This is no surprise because the congestion control methods slightly hinder the agents to move freely within the field when there is no congestion. But when the number of agents

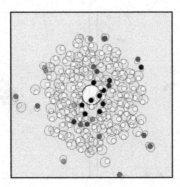

Fig. 4 Distribution of agents for a system with 150 agents after 2000 time steps using a congestion method C_D

becomes larger than 100 the performance of the system without congestion control decreases very fast. For more than 130 agents this system has the worst performance. For a medium number of agents the system with method C_W is the best. But for a large number of agents this method is not much better than a system without congestion control. For a larger number of agents (more than 210) the system with method C_D is clearly the best. Method C_P is better than the system without congestion control for more than 135 agents but it is worse than the other two methods. One reason might be that it is not so easy for the agents that have visited the service station to move away from it through the pipe because they move randomly in the considered model (and do not actively move away from the service station).

Besides the reduction of congestion, fairness for service is another important measure for the collective behavior of agents. In the considered system, e.g., the waiting times for service have to be similar. We measured the fairness of the system in two different ways. Firstly, at the end of a given time interval of length T for every agent the total amount of values that have been added to μ_i for all its visits of the service station is measured. Then the relative standard deviation (RSD) of these values for all agents has been taken as a measure for the fairness of the system (the lower the variance means the more fair the system is).

The behavior of the systems with respect to this fairness measure is shown in the left part of Figure 6. It can be seen that the system without congestion control is most fair for a small number of agents (less than 110 agents). For a larger number of agents the system with the C_D method is the best.

The second measure of fairness is defined as follows. Let $\tau(T)$ be the mean waiting time of the agents where the waiting time of an agent is defined as the length of the time interval from the time when its internal parameter (μ_i) becomes zero until the time when it reached the service area (measured over a simulation run over T time steps). Let $\sigma(T)$ be the standard deviation of these waiting times. A dimensionless measure for the fairness is then defined

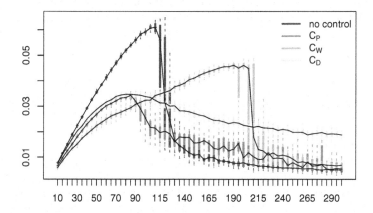

Fig. 5 Total energy consumption P_T for different number of agents measured over 10000 time steps for a system without congestion control and systems with the different congestion control methods

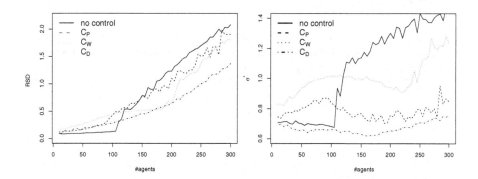

Fig. 6 Fairness for different number of agents measured over 10000 time steps for a system without congestion control and systems with the different congestion control methods; first fairness measure RSD (left), second fairness measure σ^* (right)

similarly as in [?] by $\sigma^*(T) = \sigma(T)/\tau(T)$. Note, that for this measure only the waiting times of the agents that reached the service point are considered. Hence, a congested system may still be fair, if there is only a small subset of agents that are served and these agents have similar waiting times.

The behavior of the systems with respect to the second fairness measure is shown in the right part of Figure 6. It can be seen that the system the C_D method is most fair (independently of the number of agents). For small number of agents (less than 110) the system without congestion control is the second most fair system. For larger number of agents the system with method C_P is the second best.

6 Conclusions

In this paper we have studied the problem of congestion control agent systems with ant inspired movement rules. In the studied systems the agents have to visit a service station to refill their energy storage. Three methods for self-organized congestion control have been proposed. It was shown experimentally that the proposed methods can significantly reduce the congestion and are also fair for systems with a larger number of agents. Differences between the behavior of the systems have been discussed.

Acknowledgment

This work was supported by the German Research Foundation (DFG) through the project Organisation and Control of Self-Organising Systems in Technical Compounds within SPP 1183.

References

1. M. Dorigo, V. Maniezzo, A. Colorni: Positive feedback as a search strategy. *Tech Rep.*, 91-016, Dip Elettronica, Politecnico di Milano, Italy, 1991.
2. M. Dorigo, T. Stützle: Ant Colony Optimization. MIT Press, 2004.
3. J.-L. Deneubourg, S. Goss, N. Franks, A.B. Sendova-Franks, C. Detrain, L. Chretien: The dynamics of collective sorting: Robot-like ants and ant-like robots. In *Proc. of the 1st Int. Conf on Simulation of Adaptive Behavior*, 356–363, 1991.
4. J. Handl, B. Meyer: Ant-based and swarm-based clustering. 1(2), 95-113, 2007
5. H. Schmeck: Organic Computing – A New Vision for Distributed Embedded Systems. Proc. of the Eighth IEEE International Symposium on Object-Oriented Real-Time Distributed Computing (ISORC 2005), 201-203, 2005.
6. T. Schöler, C. Müller-Schloer: An Observer/Controller Architecture for Adaptive Reconfigurable Stacks. Proceedings ARCS 05, Springer, LNCS 3432, 139-153, 2006.
7. D. Merkle, M. Middendorf, A. Scheidler: Organic Computing and Swarm Intelligence In C. Blum, M. Merkle (Eds.), Swarm Intelligence, Springer, 2008.
8. J.L. Deneubourg, A. Lioni, C. Detrain: Dynamics of Aggregation and Emergence of Cooperation. Biol. Bull., 202:262267, 2002.

9. S. Depickère, D. Fresneau, J.-L. Deneubourg. A basis for spatial and social patterns in ant species: dynamics and mechanisms of aggregation. Journal of Insect Behavior, 17(1): 81-97, 2004.

10. R.V. Sole, E. Bonabeau, J. Delgado, P. Fernandez, J. Marin: Pattern formation and optimization in army ant raids. Artificial Life, 6(3):219-226, 2000.

11. G. Theraulaz, E. Bonabeau, S.C. Nicolis, R.V. Sole, V. Fourcassie, S. Blanco, R. Fournier, J.L. Joly, P. Fernandez, A. Grimal, P. Dalle, J.L. Deneubourg: Spatial patterns in ant colonies. Proc. Natl. Acad. Sci., 99(15):9645-9649, 2002.

12. A.B. Sendova-Franks and J.V. Lent. Random walk models of worker sorting in ant colonies. Journal of Theoretical Biology, 217:255-274,2002.

13. C. Melhuish, A. B. Sendova-Franks, S. Scholes, I. Horsfield, F. Welsby: Ant-inspired sorting by robots: the importance of initial clustering. J R Soc Interface. 3(7), 235–242, 2006.

14. A. Scheidler, D. Merkle, M. Middendorf: Emergent Sorting Patterns and Individual Differences of Randomly Moving Ant Like Agents. Proc. 7th German Workshop on Artificial Life (GWAL-7), IOS Press, 11 pp., 2006.

15. M.D. Cox, G.B. Blanchard: Gaseous templates in ant nests. Journal of Theoretic Biology, 204:223-238, 2000.

16. G. Nicolas and D. Sillans: Immediate and latent effects of carbon dioxide on insects. Annual Review of Entomology, 34:97–116, 1989.

17. D. Grünbaum, Schooling as a strategy for taxis in a noisy environment. Evolutionary Ecology, 12: 503-522, 1998.

Resource-Aware Clustering of Wireless Sensor Networks Based on Division of Labor in Social Insects

Tales Heimfarth, Dalimir Orfanus and Flávio Rech Wagner

Abstract In this work concepts of division of labor in social insects and emergent self-organization are used to design a very efficient heuristic for clustering wireless sensor networks. Differently from previous approaches, we aim at creating clusters with a minimum amount of resources and good intra-cluster connectivity. Our heuristic has two steps. First, we elect the most suitable clusterheads that have the extra responsibility of leading and representing the cluster. Afterwards, the heuristic selects the respective members of the clusters. These processes are guided by a response function that determines the suitability of each node to a given task (role). For example, nodes with good connectivity and high energy level are good candidates for being clusterheads. In addition to the division of labor, we are using a positive/negative feedback mechanism to control the stimulus for attracting new members. Until having enough resources, the positive feedback acts in order to recruit new members. After gathering enough resources, the negative feedback starts to play a major role. Simulations showed that for 80% of cases the proposed heuristic could find results which are below 2.3 times the theoretical optimal solution, define as the sum of the intracommunication cost of the clusters.

1 Introduction

Wireless sensor networks (WSN) are constantly gaining popularity and attracting more research over the years. One of the reasons is a myriad of novel applications that can be implemented with them. The applications range from human-embedded

Tales Heimfarth
Federal University of Rio Grande do Sul, Brazil, e-mail: `theimfarth@inf.ufrgs.br`

Dalimir Orfanus
University of Paderborn, Germany, e-mail: `orfanus@uni-paderborn.de`

Flávio Rech Wagner
Federal University of Rio Grande do Sul, Brazil, e-mail: `flavio@inf.ufrgs.br`

Please use the following format when citing this chapter:

Heimfarth, T., Orfanus, D. and Wagner, F.R., 2008, in IFIP International Federation for Information Processing, Volume 268; *Biologically-Inspired Collaborative Computing*; Mike Hinchey, Anastasia Pagnoni, Franz J. Rammig, Hartmut Schmeck; (Boston: Springer), pp. 45–58.

sensing and ocean data monitoring to collaborative space exploration. Nevertheless, because of current hardware limitations of wireless nodes, e.g. commercial off-the-shelf sensor nodes, approaches for the management of WSN have to be designed to work using only a low amount of resources and low communication overhead.

In general, two heuristic design approaches for management of sensor networks at different levels (e.g. topology control, network layer, application) are prevalently used. The first method has in all nodes the knowledge of the (entire) network and let they manage themselves. This circumvents the need for a more advanced organization. Nevertheless, this generates overhead in terms of communication and memory at each node. Each node must, for example, maintain routes to the other nodes in the network. In large networks, the number of messages needed to maintain routing tables may cause congestion in the network and depletes the energy of the nodes. Ultimately, the need of individual self-management will generate a huge exchange of messages and overhead.

The second method identifies a subset of nodes within the network and vest them with the extra responsibility of being a leader (clusterhead) of certain nodes in their proximity. The clusterheads are normally responsible for managing communications between nodes in their own neighborhood as well as routing information to other clusterheads in other neighborhoods [1]. This creates a hierarchy in the network. Clustering in large-scale networks was proposed as a means of achieving scalability through a hierarchical approach [11]. Some examples of clustering benefits can be found at the medium access layer, where clustering helps to increase system capacity due to the promotion of the spatial reuse of the wireless channel, and at the network layer, where it helps to reduce the size of routing tables. Sensor networks and, more generally, wireless ad hoc networks largely benefit from clustering.

In this paper, we present a new heuristic that organizes a WSN into clusters. Differently from previous approaches, our proposal addresses the problem of partitioning the nodes of the network in multi-hop groups with a guaranteed minimum amount of resources q (or budget) in each one of them. This kind of clustering is useful in various scenarios. In our case, the clustering heuristic is used in the development of an efficient service distribution in our Operating System (OS).

In our OS for sensor networks, the application and OS services are distributed among different nodes and services are called remotely by the applications. Sharing the services in the network reduces the amount of resources required in each single node. An instance of the OS with all required services should be placed inside each cluster. This means that each cluster must have a minimum amount of resources.

An additional difference from our clustering heuristic to the existing ones is that we are trying to minimize the total communication cost inside the clusters. This communication cost is measured by means of a link metric that assigns a weight to each link, thus modeling the quality of the link. Moreover, the heuristic is in several aspects inspired by the behavior of various biological systems.

Our heuristic is very complex and was designed for a dynamically changing topology. In this paper we focus on the part of heuristic that deals with static topologies.

2 Related Work

In this section, a literature overview of clustering algorithms developed for sensor and ad hoc networks is presented. Some approaches were originally proposed for ad hoc networks but are also used in WSN (a subclass of ad hoc networks).

The idea of clustering is to decompose the nodes of a graph in subsets in a way that the union of the subsets contains all nodes of the graph. For each subset (or cluster), some conditions should hold.

Given a graph $G = (V, E)$ representing a communication network, where vertexes are the nodes and edges the communication links, the clustering process constructs subsets of nodes $V_i, i = 1, .., n$ where $\cup_{i=1,..,n} V_i = V$, such that each subset V_i induces a connected sub-graph of G. These vertex subsets are clusters. Ideally, the size of the clusters falls in a desired range. Moreover, for several approaches, a special vertex in each cluster is elected to represent the cluster and is called clusterhead [5].

There are several design factors concerning heuristics for cluster construction in ad hoc networks. A very important one is the maximal diameter of a cluster: when constructed as a maximum independent set (or minimum dominating set), clusters have the maximum diameter of 3. Nevertheless, we are interested in multi-hop clusters with higher diameter. Different objectives may be pursued in multi-hop clustering.

In [1], the issue of constructing the d-hop dominating set in an ad hoc network is addressed. Because of the NP-completeness of the problem for unit disk graphs, a heuristic called max-min d-cluster formation is presented. It can find good solutions with relative low communication ($O(d)$) and generalizes the dominating set problem. Nevertheless, differently from our approach, the link quality is not considered when selecting cluster members. Moreover, the size of the cluster is uncontrolled. Dense network areas result in larger clusters than sparse ones.

In [10], an algorithm for bounded size clustering based on an expanding ring search is presented. The algorithm relies on a sequence of rounds. In each round, new members are recruited for the cluster in the n-hop neighborhood. n is incremented in each round until the bound (number of nodes) of the cluster is reached. If more nodes than necessary are in the cluster, the clusterhead simply discards the excess. We compare our heuristic with a modified version of the expanding ring that guarantees clusters with a given amount of resources.

Two algorithms improving the expanding ring approach are presented in [7, 8]. They are called *Rapid* and *Persistent* clustering. As in the expanding ring, a maximum determined size (i.e. number of member nodes) for the cluster is desired. The algorithms are more efficient than the expanding ring. The cluster sizes produced should be as close as possible to the specified bound (which we will call here B) in order to limit the total number of clusters. Nevertheless, the bound should not be exceeded.

The *Rapid* heuristic uses less messages than the *Persistent* one. Nevertheless, it has a poor worst-case analytical performance. The *Persistent* heuristic persistently tries to produce a cluster of the specified bound if possible. The proposed algorithms do not violate the cluster size bound at any time. However, this bound is just given in number of nodes and there is no way to differentiate nodes. In our approach, a weight is associated to each node (representing the amount of resources of the node), and the bound is related with this weight. Moreover, the clusters in the *Rapid* and *Persistent* heuristics are always smaller than or equal to the given bound. In our approach, all clusters have at least a specified amount of resources (as can be seen in the next section).

In the *Rapid* and the *Persistent* algorithms, the clusterheads are elected in a completely random fashion, which leads to the selection of nodes that are not very suitable for the role. In our solutions we use the opposite approach: strongly connected nodes plenty of energy have a higher probability to be selected as clusterheads. Another difference is related to the links: the *Rapid* and the *Persistent* heuristics do not attempt to rank the member candidates (concerning, for example, the links) in order to select the best connected nodes to form the cluster.

A clustering algorithm where a lower and a upper bound are used to control the size of the clusters is presented in [2]. The algorithm is based on the idea of finding a rooted spanning tree of the graph (using Breadth-First-Search) and to form clusters from the subtrees that match the clustering constraints. The upper and lower bound approach tries to keep the amount of nodes in the clusters inside a specified interval. But differently from our approach, overlaps are allowed. Moreover, the link quality is also not relevant to the heuristic.

Another very important difference between all existing approaches and the one presented in this work is the fact that we try to minimize the communication overhead among all nodes inside a cluster. For that, as it will be presented in the next section, we use the smallest distance between each pair of nodes inside the clusters for the objective function. This distance is calculated by means of our combined link metric.

3 Problem Definition

In this section, a formal definition of our exact clustering problem is described.

We call our problem *minimum intracommunication-cost clustering*.

The ad hoc network is modeled as an undirected graph $G = (V, E)$, where V is the set of wireless nodes and an edge $\{u, v\} \in E$ if and only if a communication link is established between node $u \in V$ and $v \in V$. The two nodes in this case are neighbors. Each node $v \in V$ has a unique identifier (ID_v).

For each link, a weighing function assigns a positive weight. $w : E \rightarrow \mathbb{R}^+$. This weight measures the quality (or goodness) of a wireless link. We define for each edge that is not in the graph ($\{u, v\} \notin V$), that $w(u, v) = \infty$.

The quality of the link is calculated combining the following parameters: transmission success rate, received signal strength, and history of the link. The statistic-based observation of transmission success is a good indication of the future success rate. Nevertheless, it reacts slowly to changes, and at beginning there is no data to calculate its value. The received signal strength indication makes possibly quick indications, but it is not very precise. Therefore, the combined metric uses these two parameters. Moreover, in order to prioritize stable links, the history is also used. We use normalized link metrics, where 0 means a very good link and 1 a very poor one. We call the link metric *virtual distance*.

For each node, an additional weighing function r is responsible for characterizing the amount of resources available in the node. $r : E \rightarrow \mathbb{R}^+$. This models the resource capacity of the node.

The clustering process partitions the nodes into *clusters*, each one with a *cluster-head* and possibly some *ordinary nodes*. As presented in the related work section, there are several different types of clustering strategies pursuing different objectives.

In our problem, the objective is to get multihop clusters with enough resources for the OS and application processing. Moreover, the minimization of the intra-cluster communication cost is also desired.

This optimization problem is modeled as follows:

Input: A graph with weighted nodes and links (G, w, r) and a resource requirement $q \in \mathbb{R}^+$, where the sum of all node weights in each cluster must be higher or equal to q.

Constraints: For every input instance (G, w, r, q), $\mathcal{M}(G, w, r, q) = \{C_1, C_2, .., C_k | C_k$ is the kth cluster configuration$\}$, where the following properties hold

$C_k = \{c_{k1}, c_{k2}, .., c_{k(nk)}\}$ is the kth possible cluster configuration of the graph, where $k = \{1, 2, .., n\}$ (n is the number of possible configurations, nk is the number of clusters in the kth configuration, $nk = |C_k|$)

$c_{ki} = \left\{ v_{ki}^1, v_{ki}^2, .., v_{ki}^{|c_{ki}|} \right\} \in Pot(V)$ is the ith cluster of the kth configuration, where v_{ki}^j is the jth element of the cluster c_{ki}

For each configuration C_k, $k = 1, 2, .., n$, the following properties must hold:

1. $\bigcup_{i=1,2,..,nk} c_{ki} = V$ (cluster definition constraint)
2. $\bigcap_{i=1,2,..,nk} c_{ki} = \emptyset$ (no overlapping constraint)
3. Let $P(u, v) = \left\{ p_1^{(u,v)}, p_2^{(u,v)}, .., p_m^{(u,v)} \right\}$ be the set of all possible paths between nodes u and v. $p_h^{(u,v)} \in Pot(E)$ is the hth possible path where:

   $p_h^{(u,v)} = \left\{ \{u, x_1^h\}, \{x_1^h, x_2^h\}, .., \{x_{g-1}^h, x_g^h\}, \{x_g^h, v\} \right\}, x_f^h \in V, f = 1, 2, .., g, g \in \mathbb{N}$

   For each $\{u, v\} \in E \wedge u, v \in c_{ki}$, $i = 1, 2, ..., nk$, $\exists p_h^{(u,v)} \in P(u, v) | x_f^h \in c_{ki}$ for $f = 1, 2, .., g$. (Connectivity constraint)

4. $\sum_{j=1}^{|c_{ki}|} r(v_{ki}^j) \geq q$, for each $i = 1, 2, ..., nk$ (minimum amount of resources per cluster)

Costs: For every cluster configuration $C_k = \{c_{k1}, c_{k2}, .., c_{k(nk)}\} \in \mathcal{M}(G, w, r, q)$, the cost is given by:

$$cost(C_k, (G, w, r, q)) = \sum_{i=1}^{nk} \sum_{u, v \in c_{ki}} \frac{1}{2} \cdot D_{c_{ki}}(u, v) \cdot (\alpha \cdot r(u) + (1 - \alpha)) \quad (1)$$

Where $D(u, v)$ is the virtual distance between $u, v \in V$. $D_{c_{ki}}(u, v)$ is the virtual distance between u, v using just edges that are inside the cluster c_{ki}. Note that

$\forall v, u \in c_{ki}, D_{c_{ki}}(u,v) = D(u,v)$ iff the cluster c_{ki} is a convex cluster, i.e., the global shortest path between any two nodes in the clustering must use only links inside the cluster. $\alpha \in [0,1]$ controls how much the amount of resources influences the distance metric. For $\alpha = 0$, just the distances between cluster members enter into the metric; $\alpha = 1$ means that nodes with n times more resources have an n times stronger influence.

Now, we define how the virtual distance is calculated. Firstly, we introduce the cost of a path as $PCost(p_h^{(u,v)}) = w(u, x_1^h) + \sum_{f=1}^{g-1} w(x_f^h, x_{f+1}^h) + w(x_g^h, v)$. The virtual distance between u and v is the cost of the shortest path, gived by:
$D(u,v) = PCost(p_h^{(u,v)}) = \min_b \left(PCost(p_b^{(u,v)}) \right)$, for $b = 1, 2, .., m$.
The virtual distance using only nodes inside the cluster is defined by:

$$D_{c_{ki}}(u,v) = PCost(p_h^{(u,v)}), \text{ where } p_h^{(u,v)} \in P(u,v) | x_f^h \in c_{ki} \text{ and } PCost(p_h^{(u,v)}) = \min_b \left(PCost(p_b^{(u,v)}) \right), \text{ for } b = 1, 2, .., m$$

Goal: *Minimum*, i.e. $\min_k \{ cost(C_k, (G, w, r)), \text{ for } k = 1, 2, .., n \}$

The *minimum intracommunication-cost clustering* is an NP-complete problem. The proof can be performed by means of reducing the *partition problem* to our clustering problem (*partition problem* $\leq_p$ *minimum intracommunication-cost clustering*). The complete proof can be seen in [6].

4 The Emergent Clustering Heuristic

Our heuristic cluster construction process consists of two subparts: (1) The clusterhead election, responsible for selecting a subset of nodes and vesting them with the extra responsibility of leading and representing the cluster; (2) The membership selection, responsible for selecting the members of a cluster. Both subparts use behaviors and principles observed in the nature.

Clusterhead election (Section 4.2) is inspired by division of labor and task allocation in swarms of social insects, described in detail by Bonabeau et al. [3]. The possible tasks (or roles) that a node can assume are:

Clusterhead (CH):   The clusterhead nodes are the representatives of the clusters. The identification of the cluster is given by the clusterhead. Moreover, special tasks are assigned to the clusterhead. Once the clusterhead is not present in a cluster anymore, the cluster ends its existence.

Member (Me):   The members of the cluster are the nodes that have decided which cluster they belong to.

Ordinary Node (Not member, Nm):   Nodes that neither decide to enter a cluster nor become clusterhead.

In the case of membership selection (Section 4.3), we are combining division of labor with the concept of emergence of self-organization. Self-organizing systems acquire structure by themselves and are normally composed by a large number of locally interacting components. [4] presents two basic modes of interaction among the components: positive and negative feedback. Our emergent clustering heuristic is specifically inspired on the behavior of the male bluegill sunfish (*Lepomis macrochirus*), which uses for nesting these two modes of interaction.

Positive feedback can be simplified as the behavioral rule "I nest where others nest". The nesting pattern appears in a large lake with an initial homogeneous structure due to the amplification of fluctuations: if the density of bluegills is sufficient, through a random process, several nesting sites will be occasionally close enough to provide a sufficient attraction that stimulates even more bluegills to nest nearby. This random pattern of nest sites now becomes unstable and a cluster of nest sites will grow. A process like this, with positive feedback, is also called an autocatalytic process.

The negative feedback is responsible for controlling and shaping the system in a particular pattern. Without it, a potential destructive explosion may be easily reached. The feedback can be rephrased as "I nest where others nest, unless the area is overcrowded". Physical constraints like depletion of the building blocks can be also included in the negative feedback.

As the result of the interplay of these modes of interaction, a nice-shaped cluster of nests emerges at the bottom of a lake. This happens without any central control or blueprint, exactly like in our heuristic.

4.1 Overview of the Approach

The first task of the heuristic is to elect the clusterheads of the network using the response function $T_{\theta_{ch_v}}$:

Nonmember $\rightarrow$ Clusterhead: The response threshold function $T_{\theta_{ch_v}}$ returns the probability of a nonmember v to become a clusterhead. The function is responsible for modeling the emergence of clusterheads in areas of the WSN where no clustering is already taking place.

A clusterhead is now a unitary cluster with some resource ($R_i = r(v)$, v is the clusterhead of cluster i). When a clusterhead is elected in some part of the network, as a consequence of missing resources it starts to "attract" new members with help of the response function $T_{\theta_{recr_{v,i}}}$:

Nonmember $\rightarrow$ Member of x: The response function (recruitment function, $T_{\theta_{recr_{v,i}}}$) models the recruiting of new cluster members through a positive feedback process. It provides the probability that node v will enter into the cluster $ID = i$.

The idea is that a cluster incrementally grows until it achieves at least the requirement q of resources. The intensity of the attraction force (and consequently the stimulus to enter into the cluster) is regulated by the amount of resources already in the cluster. A growing cluster exercises an attraction force to the nodes that are in the vicinity. This attraction force is expressed by a higher stimulus s in the $T_{\theta_{recr_{v,i}}}$ response function (positive feedback). Then, when a cluster attracts nodes that bring enough resources, the attraction force becomes much smaller (negative feedback).

4.2 Clusterhead Election

As we mentioned before, the clusterhead has an extra responsibility of representing the cluster and leading the selection of members. Nodes have different predispositions to be a clusterhead, i.e. they have distinct connectivity and distinct amounts of energy. It is obvious that the clusterhead should have good connectivity to other nodes and enough energy to cover the extra activity due to the leadership (build-up and maintenance of a cluster). The opposite is also true: nodes with poor connectivity and an almost depleted source of energy are not good candidates. This concept is derived from the division of labor of social insects. Instead of having just a certain number of fixed morphology agents (like the *majors* and *minors* in the *Pheidole* genus), we have here the complete spectrum of nodes: from nodes very capable of assuming the clusterhead role to nodes not suitable at all for this task. We model the probability of node v to become a clusterhead with the response function

$$T_{\theta_{CH_v}}(s_{CH_v}) = \frac{s_{CH_v}^{\beta}}{s_{CH_v}^{\beta}+\theta_{CH_v}^{\beta}}.$$

The fitness of the node to the role of clusterhead is modeled in the response function with the threshold (θ_{CH}). A small threshold means that the node is very suitable to be a clusterhead. Parameter s_{CH} models the stimulus to become a clusterhead. For a given threshold, a high stimulus increases the probability of the node to become a clusterhead.

The definition of threshold is in Equation 2.

$$\theta_{CH_v} = k_1\left(\frac{\sum_{u \in Ngb_{Nm}(v)} w(u,v)}{|Ngb_{Nm}(v)|}\right) + k_2(1-E_v) + k_3\left(1 - min\left(1, \frac{|Ngb_{Nm}(v)|}{Max\_Neighb}\right)\right) \quad (2)$$

Where $Ngb_{Nm}(v)$ is the set of all neighbor nodes which are in nonmember state, $w(u,v)$ measures the quality of the link between two nodes, and E_v describes the energy level of the node.

As we said before, factors that influence the threshold are good connectivity (the first and the third term) and amount of energy (the second term). Each factor has a different importance for the overall threshold, which is captured with weights (k_1, k_2, k_3). Weights range from 0.0 to 1.0, and the sum of them is 1.0.

The stimulus function is given by $s_{CH_v} = k_1 \frac{t_{elapsed}}{t_{required}} + k_2\left(1 - \frac{|Ngb_{Me}(v)|+|Ngb_{CH}(v)|}{|Ngb(v)|}\right)$.

Where $t_{elapsed}$ is the elapsed time since the clustering heuristic has started and $t_{required}$ is the maximum running time of the algorithm. $Ngb(v)$ is the set of all neighbors of the node v, $Ngb_{Me}(v)$ is the set of nodes in member state and is subset of $Ngb(v)$, the same for $Ngb_{CH}(v)$.

As we can see, there are two factors that stimulates a node to become a cluster-head: (1) nodes that for a long time did not belong to any cluster (first term); and (2) nodes without clusters in the vicinity (second term).

Based on the response function presented, each node periodically tests whether it should become a clusterhead. Initially, all nodes are nonmembers. With time, clusterheads will emerge and attract other nodes to be a member of their clusters.

If a clusterhead, after a certain number of attempts, could not keep the requirement q of resources per cluster, then the (incomplete) cluster will cease its existence and the current members will be free to join other existing clusters.

4.3 Member Selection

Once clusterheads emerge, they start to send messages to attract new members. Each nonmember that receives this message will evaluate its probability of assuming the task of member of the cluster using the response function: $T_{\theta_{recr_{v,i}}} = \frac{s_{recr_{v,i}}}{s_{recr_{v,i}} + \theta_{recr_{v,i}}}$.

Where the threshold and the stimulus have the following meaning:

Threshold $\theta_{recr_{v,i}}$: measures how connected the node v is to the cluster i. A small value means high suitability to be a member.

Stimulus $s_{recr_{v,i}}$: represents the volition of a cluster to attract new members. Here the positive and negative feedback act.

The threshold function for node v is defined by:

$$\theta_{recr_{v,i}} = k_1 \cdot D_i^v + k_2 \cdot min \left\{ \frac{D(v, Clusterhead_i)}{Max\_dist}, 1 \right\} + k_3 \cdot min \left\{ \frac{Cn_i^v}{Max\_connect}, 1 \right\} + k_4 \cdot \frac{r(v)}{q} \quad (3)$$

Where D_i^v is the distance to the nearest member of the cluster i and $D(v, Clusterhead_i)$ is the distance to the clusterhead. $Cn_i^b = \sum_{e \in \{Ngb(b) \cap c_i\}} (1 - w(b, e))$ measures the connectivity to the neighbors that are already in the cluster using the link metric w.

The first factor that influences the threshold (first term) reduces the distance among members of the cluster. The factor that influences the shape of the cluster is captured in the second term. Advantage is given to flat configurations (small cluster diameter).

The selection of nodes that are well connected to members of the cluster increases the probability of reducing the cluster cost. This idea is reflected in the third term.

The fourth term covers the idea that nodes with higher resource availability will potentially reduce the cost of the cluster because they reduce the necessity of taking additional nodes.

The stimulus of a node to belong to the cluster i is given by $s_{recr_{v,i}} = k \cdot (p(R_i) \cdot g(R_i))$.

If two clusters are trying at the same time to attract the node, this equation is used with the higher stimulus. The stimulus is the combination of positive and negative feedbacks.

Aggregation Through Positive Feedback

Positive feedback is used to control the stimulus of neighboring nodes to enter a determined cluster. It is performed by considering the attraction force (or stimulus in the response function) to be proportional to the amount of resources R_i of the cluster i plus some bias, i.e., $p(R_i) = k_1 + k_2 \cdot R_i$. This equation denotes the relationship between the amount of resources and the "force" (that is reflected in the stimulus) to attract new nodes to the cluster.

Creating Structure Through Negative Feedback

The negative feedback is responsible for "controlling" the explosive nature of the positive feedback and to shape the emergent structures in the self-organizing process. In our case, we use Equation $g(R_i) = 1 - \left(\frac{R_i}{k_1 \cdot q} \right)^{\beta}$ as negative feedback.

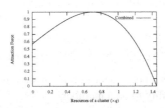

Fig. 1 Resulting attraction force after combination of the positive and the negative feedback

It is important to remark that the negative feedback in our case controls how much the positive feedback takes effect, i.e., the result stimulus is given by the multiplication of the feedbacks, a fact that is shown in Figure 1.

4.4 Cluster Construction Process

In this section we will present the steps performed by the heuristic to build the clusters based on the concepts presented in the previous sections.

At the beginning, there is no cluster in the network. Every node tests periodically whether it should become clusterhead (using the response function $T_{\theta_{chv}}$). An information flow based on beacons is used to provide the nodes with the necessary knowledge for the response function.

When the node v decides to become clusterhead, a new cluster (we call it cluster i, $i = clusterID$) comes into existence. Initially, this cluster has the resource $R_i = r(v)$.

Now, it starts to broadcast to the neighborhood periodically its current resource state (R_i). The message is called clusteringForward. The basic function of the clusteringForward message is to inform all members of the cluster and nearby neighbors the actual amount of resources of the cluster. This is used by the nodes to calculate the current attraction force of the cluster. The clusteringForward message is forwarded by the members of the cluster until arriving at nodes outside the cluster. During this phase, a spanning tree having the clusterhead as root is generated. Nodes outside the cluster that receive a clusteringForward message will generate the clusteringBackward message that travels back to the clusterhead, gathering information about nodes with intention to enter or leave the cluster. Each node that is not a leaf of the spanning tree waits until receiving the clusteringBackward message from its children before sending a fused clusteringBackward message to its own parent.

We will call this process of sending the clusteringForward message and gathering information through the clusteringBackward message a *cluster construction round*.

As already said, the cluster construction round is started by the clusteringForward message issued by the clusterhead. When receiving this message, a node u stores it temporarily in order to select the message with the smallest link metric to the clusterhead. This is used to build a good spanning tree with the clusterhead as root.

The way of responding to the incoming message varies depending on the current status of the node u:

Node u is not a member of cluster i: The first action of the node is to determine whether it should enter the cluster i. This is done using the response function $T_{\theta_{recr_{v,i}}}$ (recruitment function) to evaluate whether the node u wishes to enter the cluster (recruitment function). This response function uses the connectivity to the cluster as threshold (good connected nodes have less threshold to enter the cluster), and the stimulus is given by the combination of the positive/negative feedback presented in Section 4.3. If the test of the recruitment function returns positive, the clusteringBackward message will carry the membership intention of the node u. The next *clusteringForward* message will confirm (or not, if the cluster is overcrowded) the acceptance of the node u in the cluster.

Node u is a member of cluster i: The node will test whether it should leave the cluster using the response function $T_{\theta_{leave_y}}$. If the test returns negative, the node just retransmits (forwards) the message clusteringForward in order to continue the construction of the spanning tree. If the node is willing to leave the cluster (because its connection is getting loose), it also forwards the clusteringForward message, but indicating this intention of leaving the cluster. This will force previous children to select another parent because this node is going to be disconnected from the cluster. If they could not find another parent, they must also disconnect themselves from the cluster.

The clusteringBackward message is used to inform the clusterhead about nodes with intention to enter the cluster and nodes willing to leave. Moreover, the *id* of all members of the cluster is collected in this message. Therefore, the clusterhead can re-check the complete membership of the cluster to see if some node has for example disappeared due to failure or a drastic topology change.

When the clusterhead receives the clusteringBackward message from all its direct children, it can decide which nodes that are willing to enter the cluster will be accepted. This decision is based on their thresholds to enter the cluster: nodes with less threshold have higher priority.

It is important to state here that after the cluster is complete, the clusterhead ceases to start new rounds. When some member of the cluster detects a large topology change, the clusterhead is informed and a new round is started to re-check the complete cluster (reactive response to topology changes).

An example of the cluster construction round is shown in Figure 2.

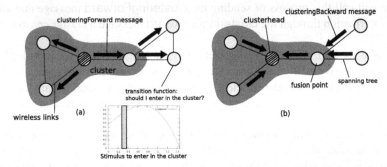

Fig. 2 Example of cluster construction round. (a) Clusterhead starts the round sending the message clusteringBackward with the current amount of resources of the cluster. (b) When arriving at nodes outside the cluster, they decide whether they are willing to join the cluster. This information is sent back using the clusterBackward message

The first purpose of the positive/negative feedback is to reduce the amount of information aggregated in the *clusteringBackward* message. Nodes badly connected to the cluster will decide not to enter the cluster, thus reducing the amount of information that the *clusteringBackward* message must carry.

The second purpose of the positive/negative feedback mechanism is to control the competition among neighboring clusters and belongs to the dynamic part of our heuristic (which is not the main focus of this paper). The feedback curves are designed in such a way that an already formed cluster may just loose some members till the q limit is achieved, because, when this limit is achieved, the desire to attract new members is at maximum. In the same way, if there are two clusters under construction, this method avoids that one cluster steals members from the other one, reaching the state where no cluster has fulfilled its requirement on resources.

5 Results

We implemented our emergent clustering heuristic using an event-based wireless ad hoc simulator called ShoX [9]. Some parts of the heuristic were also implemented in the specification and modeling language AsmL. As input, we generated 35

instances of the problem with 16 nodes in a field of 50m by 50m. These instances
were generated by random selection of the node positions.

We used the received signal strength (RSSI) for the free space model with
isotropic point sources as link metric. We decided to test our heuristic for dense
networks, therefore the radio range was 70m, covering the complete field.

In this paper, we evaluate our heuristic for static networks and for networks of
homogeneous devices (i.e., each node has a unit of resource). We run our algorithm,
let it converge to a stable configuration, and compare the heuristic result with the
optimal one and also with an existing heuristic called expanding ring [10]. In order
to calculate the optimal result, we model our *minimum intracommunication-cost
clustering* as an integer linear programming model and, for each simulated instance,
we find the optimal solution for that configuration with the lp_solve program.

Figure 3 shows results of performed experiments. In average, the cost of our
heuristic was 1.98 times higher than the optimal solution and the expanding ring
was about 4.29 times higher. To run the simulation of the complete network, it took
10 seconds while the optimal solution needed more than 10 hours in Intel Core Duo
2.7 GHz computers.

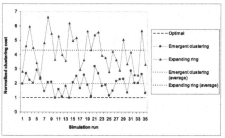

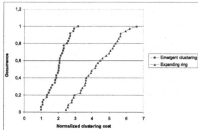

Fig. 3 Normalized results of performed experiments

Fig. 4 Cumulative distribution of normalized
simulation results

In Figure 4, the cumulative distribution of the normalized results can be seen. It is
possible to notice that for more than 80% of all simulations, the emergent clustering
heuristic could find results that were below 2.3 times the optimal one. In the case of
the expanding ring, results were below 5.2 times the optimal one.

6 Conclusion

In this paper, we introduce a useful clustering problem and develop an efficient
heuristic inspired by biological systems to solve it. The heuristic has two parts: the
clusterhead election, which is responsible for selecting a subset of nodes and vesting them with extra responsibility of representation of the cluster, and membership
selection, which is responsible for selecting members in order to fulfill the resource
requirements of each cluster.

The selection of the task for a node is based on its suitability for that task. In the same way that ants with different morphology have tendency to perform different tasks, different nodes have different probabilities of assuming the clusterhead or cluster member roles. This concept is combined with a positive/negative feedback stimulus, which is responsible to shape the size and form of the cluster.

The results of the simulations show that the heuristic performs well, with cost in average just 1.98 times the optimal one. This was achieved in a distributed manner and using only locally available information to make decisions. This makes this heuristic suitable for ad hoc networks with resource-constrained devices or sensor networks.

The results obtained here re-enforces our confidence that methods found in nature can be successfully transferred to computer systems.

In the future, we plan to simulate the heuristic in networks with moderate topology changes, evaluating our approach with dynamic scenarios.

References

1. A. D. Amis, R. Prakash, T. H. P. Vuong, and D. T. Huynh. Max-min d-cluster formation in wireless ad hoc networks. In *INFOCOM 2000. Nineteenth Annual Joint Conference of the IEEE Computer and Communications Societies. Proceedings. IEEE*, volume 1, pages 32–41vol.1, 26-30 March 2000.
2. S. Bannerjee and S. Khuller. A clustering scheme for hierarchical control in wireless networks. In *Proceedings of the IEEE INFOCOM*, Anchorage, AK, April 2001.
3. Eric Bonabeau, Marco Dorigo, and Guy Theraulaz. *Swarm Intelligence: From Natural to Artificial Systems*. Oxford University Press, Santa Fe Institute Studies in the Sciences of Complexity, New York, NY, 1999.
4. Scott Camazine, Jean-Louis Deneubourg, Nigel R. Franks, James Sneyd, Guy Theraulaz, and Eric Bonabeau. *Self-Organization in Biological Systems*. University Presses of CA, 2003.
5. Y. Chen, A. Liestman, and J. Liu. Clustering algorithms for ad hoc wireless networks. In Ad Hoc and Sensor Networks, 2004.
6. Tales Heimfarth. *Biologically Inspired Methods for Organizing Distributed Services on Sensor Networks*. PhD thesis, University of Paderborn, 2008.
7. R. Krishnan and D. Starobinski. Message-efficient self-organization of wireless sensor networks. In *Proceedings of IEEE Wireless Communications and Networking Conference (WCNC)*, New Orleans, USA, March 2003.
8. Rajesh Krishnan and David Starobinski. Efficient clustering algorithms for self-organizing wireless sensor networks. In *Ad Hoc Networks*, volume 4, pages 36–59, January 2006.
9. Johannes Lessmann, Tales Heimfarth, and Peter Janacik. Shox: An easy to use simulation platform for wireless networks. In *Proceedings of The 10th International Conference on Computer Modelling and Simulation*, Cambridge, England, April 2008.
10. C. V. Ramamoorthy, A. Bhide, and J. Srivastava. Reliable clustering techniques for large, mobile packet radio networks. In *Proceedings of the 6th Annual Joint Conference of the IEEE Computer and Communications Societies (INFOCOM 87)*, San Francisco, USA, April 1987.
11. Ivan Stojmenovic, editor. *Handbook of Sensor Networks*. John Wiley and Sons Inc, 2005.

Self-stabilizing Automata[*]

Torben Weis and Arno Wacker

Abstract Biological systems are known to be probabilistically self-stabilizing, i.e. with a high probability they can reach a stable state from any initial state. This property is very important to computer-based systems, too. However, building self-stabilizing systems is still very difficult. Proving that any given implementation is in fact self-stabilizing is even harder. Nature has a big advantage: Any living being must eventually die and limited energy limits the harm that an error can have on the system. This greatly simplifies the realization of self-stabilization. To transfer this concept to computer-based systems, we propose to modify the computational model on which software is currently being built. We introduce energy-awareness in Turing Machines (TMs). This will guarantee that any TM program that is correct in the absence of errors is at the same time self-stabilizing in the presence of errors.

1 Introduction

Today's software often assumes that errors do not occur. Better software designers define at least an error model. For example, they assume network errors but no memory errors. If errors occur which do not fit in the error model, the error is neither detected nor corrected. If the error fits in the error model, it can be detected and by default the application is stopped. In the best case, the error is detected and corrected.

Biological systems are different. They do not feature any error detection and they don't throw exceptions. In the physical world all possible states are allowed

Torben Weis
University of Duisburg-Essen, Duisburg, Germany e-mail: torben.weis@uni-duisburg-essen.de

Arno Wacker
University of Duisburg-Essen, Duisburg, Germany e-mail: arno.wacker@uni-duisburg-essen.de

[*] This work has been supported by the DFG SPP Organic Computing

Please use the following format when citing this chapter:

Weis, T. and Wacker, A., 2008, in IFIP International Federation for Information Processing, Volume 268; *Biologically-Inspired Collaborative Computing*; Mike Hinchey, Anastasia Pagnoni, Franz J. Rammig, Hartmut Schmeck; (Boston: Springer), pp. 59–69.

and biological systems have the tendency of developing from most initial states to a set of preferred states. In these preferred states, living beings exist and reproduce themselves.

In computer science we divide all possible states in valid and invalid states. Furthermore, computer programs assume a precisely defined initial state. Any state that can be reached without errors from this initial state is then called a valid state. A system is called self-stabilizing if it transits from any invalid state to a valid state in a constant amount of time.

In the related work section we discuss algorithms which are known to be self-stabilizing. However, the development of such algorithms and the proof of their self-stabilization property require much time and expertise. In most commercial applications this is simply too expensive. Furthermore, real-life systems are very complex which renders theoretical proofs next to impossible.

Obviously, there seems to be a major difference between software development and biological evolution. In biology, evolution is constantly changing the genes and up to now every known genetic program either terminated (i.e. the species died) or it could reach a stable state where it continuously reproduces itself. Furthermore, no species has been accidentally created that happened to eat up the universe or bring it to a halt. Unfortunately, this is what we have to expect if we apply arbitrary mutations on software programs. We will most likely end up with a program that neither terminates nor reaches a valid state. Even worse, it will consume all CPU time, disable all interrupts and lock the computer.

We claim that the problem is the automaton which executes the programs. Biological systems are being executed on a physics engine which follows a set of fundamental laws. Computer systems are being executed on machines which are a Turing Machine or some equivalent. If we want to develop computer software inspired by biology, we must first fix the computational model, i.e. the machine on which the software is executed. In this paper we present a modified Turing Machine which takes fundamental laws of physics into consideration. The result is that applications running on this machine are automatically self-stabilizing.

2 Application Scenario

Biologically inspired software is not necessarily the best approach for all application scenarios. In some application domains errors are not part of the normal operation. Hence, if an error is detected, the system is stopped and the administrator must fix it. For example, office and enterprise applications belong to this category.

Applications dealing with sensors and actuators are different. Temporary sensing errors or temporarily broken actuators belong to the normal mode of operation and there is no system administrator available for fixing every possible problem. Thus, self-stabilizing systems are preferable, because they can autonomously recover from a wide range of errors without any intervention by a system administrator.

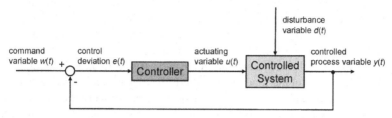

Figure 1 Control Loop

The application scenario for our research consists of a wide range of control applications. The structure of control applications is shown in Figure 1. The controlled system consists of actuators and sensors and interacts with the physical world. The controller is implemented in software and communicates with the controlled system by exchanging messages. Based on the sensor input and its internal state the controller sends commands to the actuators. The behaviour of the actuators in turn influences the sensors. Thus, the system is a software controlled feedback loop. Errors or disturbances can be introduced on the controlled system. If the controller is self-stabilizing, it can recover from any temporary sensing error and from any temporarily broken actuator in constant time. Thus, for every system exists a constant time t such that the entire system reaches a consistent state in no more than t seconds after all temporary errors are gone.

The aim of our research is to create a software development process and tools for building self-stabilizing controller software. In addition, we believe that the results of our work can be applied to other application scenarios, for example in the area of pervasive and ubiquitous computing.

3 Physics versus Turing Machines

The Turing-Church-Thesis claims that every effectively computable function can be regarded as computable under the definition of the Turing Machine. It does by no way claim the converse. Not every function that can be computed by a Turing Machine can be computed by a physical machine. The typical argument is that a TM has an infinite tape whereas all physical machines are limited. However, the difference between physics and Turing Machines is not only a matter of tape length.

Our argumentation is that Turing Machines do not obey the second law of thermodynamics, which states that "the total entropy of any isolated thermodynamic system tends to increase over time, approaching a maximum value." In contrast, a Turing Machine can work until eternity on a program that continuously increases the entropy of the tape. Even if we could build a physical computer with infinite amount of memory, it would still not be Turing equivalent because it must obey the second law of thermodynamics.

As a consequence of this observation, we modified the Turing Machine. The first law of thermodynamics states that "in a closed system energy can neither be created nor can it disappear. It can only be transformed in other kinds of energy" (e.g. thermal energy or work force). Thus, we had to introduce a concept that is comparable to thermal energy and work force and a transformation between both of them. The second law of thermodynamics limits the transformation between thermal energy and work. It implies that it is impossible to construct a process that translates thermal energy lossless to work force. This is often expressed as: "Perpetuum Mobili cannot exist". Thus, our machine must have a way of transforming thermal energy to work force in a non-lossless way only.

We do not want to overstress the parallels to physics. However, the laws of thermodynamics have been the starting point of our approach and inspired our machine. Furthermore, these laws describe very well that there are some major differences between the computational model used by biology (the laws of physics) and the computation model of computer systems (Turing Machines).

4 Energy-aware Turing Machines

In our approach we assume that the read/write head of the TM has a certain thermal energy. The tape has the thermal energy 0 and no symbols exist on the tape initially, i.e. the head is hot and the tape is cold. The thermal energy of the read/write head can be transformed into work force and it can be transferred to the tape and its symbols. A read/write head can perform three kinds of work. It can move, read, or write. Performing any of these actions affects the thermal energies. We assume that the tape is huge, i.e. we can transfer much thermal energy to the tape without changing its temperature significantly. The symbols are in contrast tiny. Little energy transfer is required to heat them up. The head is supposed to be much larger than the symbols but small compared to the tape. Size matters because it determines how much energy is required to change the temperature of an entity.

In our machine the head is moving upwards and downwards. Moving the head upwards transforms thermal energy of the head into potential energy. Moving the head downwards transforms potential energy back into thermal energy. Because of the second law of thermodynamics this transformation process must not be lossless. Therefore, during each movement the read/write head heats up the tape, i.e. transfers the thermal energy $\Delta E > 0$ from the head to the tape (see Figure 2). As a result no energy is ever lost or created and a perpetuum mobile (i.e. a head that is moving forever) is impossible because more and more energy is transferred to the tape.

Over time the symbols exchange thermal energy with the tape. Thus, the symbols cool down. Eventually the tape and all symbols will have the same temperature and can no longer exchange thermal energy. A symbol which has the same temperature as the tape is no longer readable and disappears. Thus, the head must always write symbols which are at least as hot as the tape. If this is no longer possible, no symbols can be written any more.

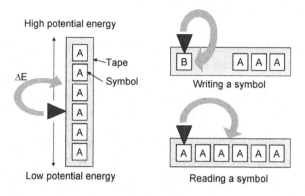

Figure 2 Operations of an energy-aware Turing Machine

Writing a symbol to the tape transfers thermal energy to the symbol until head and symbol have the same temperature. This implies that hot heads write hot symbols and cooler heads write cooler symbols. This is in line with the second law of thermodynamics because the thermal energy flows from a warm entity (the head) to a cooler entity (the symbol) until both having the same temperature.

If the head is reading a symbol, it cools down until the head and symbol have the same temperature. The thermal energy lost by the head in this process is transferred to the tape. Thus, the cooler the symbol that is being read, the cooler the head becomes while reading it. This implies that the head cannot read a cold symbol and write a hot symbol afterwards. The energy of a freshly written symbol is always higher than the energy of any symbol written afterwards.

The head cannot move until eternity because it constantly looses thermal energy to the tape. Eventually the only possible movement is downwards because this is the position with the lowest potential energy and the head cannot spend more thermal energy on moving upwards. Eventually, the machine will fall back to its initial state. The head falls down to the lowest position and all symbols disappear once they reach the same temperature as the tape.

The amount of energy that is lost (i.e. transfered between head or symbols and tape) determines the stabilization time. This energy loss must always be higher than 0 to avoid a perpetuum mobile. The higher the loss, the faster will the head return to its initial position and the faster will erroneous symbols disappear. On the other hand, high energy losses mean that the system will forget fast, i.e. its view on past sensor values is very narrow, because old values disappear very quickly.

The machine presented so far is in line with the rules of thermodynamics. However, it is not very useful yet. The machine has simply a limited time for execution. After this time all parts of the machine have the same temperature and the machine resets to its initial state. In the next chapter we will therefore extend our machine to read sensor values and control actuators.

5 Sensors and Actuators

The rules of thermodynamics cited so far apply to closed systems. However, if we allow our machine to receive sensor data from outside the machine and to control actuators outside the machine, the system is no longer closed. We assume that the machine increases its thermal energy when it receives data (from a sensor) and emits thermal energy when it sends data (to actuators). As long as sensors continuously send more data, the machine does not necessarily cool down to the point where it falls back to its initial state (head at the lowest position and no symbols on the tape).

Our machine is in fact a three tape Turing Machine as shown in Figure 3. The middle tape is the working tape. The first one is the input tape and the last one the output tape. A sensor is sending input data as a finite sequence of symbols. If the machine is in a receiving state (which is the case when the head is at the initial position) these input symbols are copied on the input tape. The head moves along all received symbols and they transfer thermal energy to the head. Thus, the energy level of the machine is increased and it can read and process the data.

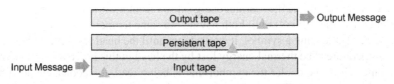

Figure 3 3-Tape Turing Machine

The machine can write symbols on the output tape. When the machine falls back to the initial position, it can receive new input and it sends its output. The symbols on the output tape are sent to actuators. By doing so these symbols disappear from the output tape because their thermal energy is sent to the actuators. The machine looses energy by sending.

The energy limits the number of symbols the machine can send. This shows already one great advantage of this machine. The damage its output can cause is limited by the input it receives. As long as sensors are sending only at a limited frequency and with limited thermal energy, the machine cannot run berserk by flooding the system with bogus symbols. For example, accidental distributed denial of service attacks are impossible because of energy restrictions. This argument holds for all programs executed by this machine.

6 Self-stabilization

The inability of our machine to store data forever (symbols loose energy to the tape) and its inability to amplify the energy of stored data (cannot read a cold symbol and write a hot one) is the key to its self-stabilization property. We assume that the machine is executing a program that is correct if no errors occur. Possible errors are:

- Tape symbols are accidentally added, removed, or modified
- The head position is accidentally changed
- Sensor data is lost, duplicated, modified or its ordering is changed
- Too much energy is sent to the machine
- The machine has been loosing to much energy

If anything is wrong with the current state of the machine, then a certain thermal energy is attached to this wrong information. It could be a wrong symbol or a wrong head position. After a constant time the head must move back to its initial position and after a constant time all symbols have cooled down and disappear. All information disappears after a constant time and it cannot be refreshed, i.e. its energy cannot be amplified. Furthermore, the energy of derived information cannot be higher than the energy of the source information, because the head cannot write a symbol that is hotter than any other symbol read or written before. After a constant time it does not matter whether some information was wrong or not because the information itself and all information derived from it disappears. What remains has necessarily a high energy level and is therefore fresh information that is in no way dependent on the wrong information and therefore correct. The only source for fresh information is new sensor readings. Thus, after a constant time all erroneous information is gone and only fresh and correct information remains in the machine.

Our machine greatly simplifies the development of self-stabilizing algorithms. The developer does not have to prove that his algorithm can recover from every possible error. He must prove that the algorithm executes correctly on our machine in the absence of errors. If this is the case, the self-stabilization property is guaranteed. This is a great advantage over current coding techniques where the self-stabilization property requires a manual proof.

However, the proof of correctness is now a bit more complex than with normal Turing Machines. The proof of correctness must take the energy transfer into account. In addition, no sane programmer will develop an algorithm for execution on a Turing Machine. Therefore, we are working on a model-driven development approach [1]. The developer describes his program on a very high-level programming language [2]. This program is then automatically translated into a program for our machine. This program can now be tested on a simulated machine. Testing is of course no proof of correctness, but in practice testing is easier to do than correctness proofs.

A disadvantage of our machine is that it is even less efficient than normal Turing Machines. The CPU must calculate the energies whenever the head moves, reads, or writes. Thus, it is no platform on which you would execute a word processor or enterprise applications. However, a machine that is inherently forgetful is not useable for this kind of applications anyway.

For control applications, however, our machine is well suited. A constant performance factor is often tolerable. Furthermore, the forgetfulness of our machine is no problem here. When a control application receives input, it calculates its output

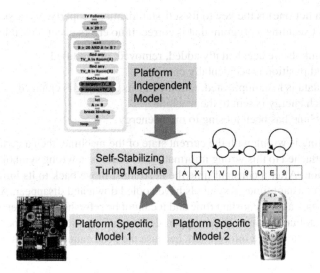

Figure 4 Development process

based on the new input and a fixed amount of previous inputs. Very old inputs are not required, which is good, because our machine has already forgotten these old inputs and every data derived from them.

7 Implementation and Simplification

So far the energy-aware Turing Machine is a theoretical concept. To turn it into something useful, we must execute the energy-aware Turing machine programs on a real machine. Although it would be an interesting challenge to build a thermo-dynamic machine which adheres to our formal specification, this is of course not a practical thing to do. Instead, we see two possible options: A software solution and a silicon hardware solution.

In the first case, the programs are executed in a simulated energy-aware Turing Machine. This is the easiest thing to do and ensures that the program is self-stabilizing as long as the underlying software is stable. However, if the simulator is not working correctly or if the operating system crashes then the entire system will not be self-stabilizing. The more radical approach is to build hardware in silicon which behaves like an energy-aware Turing Machine. In this case there is no software layer (simulator or operating system) that could fail. Such a system would really be self-stabilizing. Since our expertise is not in chip design, we are working with the simulator approach currently.

In both cases, the energy-aware Turing Machine is hard to implement because it needs much computation to calculate the energies as floating point numbers. Therefore, we simplified the energy-aware Turing Machine without sacrificing its self-stabilization properties. First of all, energy is quantized to represent energy levels as integers. Furthermore, the potential energy of the read/write head can be ignored. It has only been introduced to make sure that the head falls back to the initial position once it lost too much thermal energy. In an implementation, we simply check after each step whether the head energy level is too low and move it back to the initial position.

The symbols are constantly transferring energy to the cooler tape until they disappear. We implement this by storing for each symbol the step in which it has been written and its energy level at this time. If the head later on reads the symbol, we subtract one from the energy level for every step that happened in between. If the resulting energy level is 0 or below, the symbol is erased.

It could happen that a memory error changes the step or energy level stored for some symbol. This does not harm the self-stabilization property as long as all of these values are expressed as numbers with a fixed amount of bits. Thus, if a memory error accidentally increases the energy level of a symbol, the additional energy is limited by the amount of bits. After a constant time this energy has been transferred to the tape, the symbol disappears, and the system can stabilize again. The fewer bits we use, the shorter is the self-stabilization time. However, the forgetfulness of the machine increases when the number of bits decreases. The best number of bits to use is therefore a trade-off between stabilization time and forgetfulness.

8 Related Work

In our approach we are extending a three-tape Turing Machine which is known as Persistent Turing Machine (PTM). PTMs [3, 4] are a minimal extension of Turing Machines (TMs) [5] that express interactive behaviour providing a natural model for sequential interactive computing. A PTM has three tapes: a read-only input tape, a write-only output tape, and a persistent working tape which is preserved among interactions, i.e., among successive computations of the PTM.

Self-stabilizing algorithms have been introduced by Dijkstra in 1974 in his seminal paper on a self-stabilizing token passing algorithm [6]. A self-stabilizing system recovers from any transient fault within a bounded number of steps [7] provided that no further fault occurs until the system is stable again. The maximum number of steps required to bring the system back into a legitimate state is called stabilization time. Self-stabilization is usually proven by showing that the system satisfies convergence (started from an arbitrary state it reaches a legitimate state within a bounded number of steps) and closure (once the system has reached a legitimate state, it stays in the set of legitimate states) if no faults occur.

Many self-stabilizing algorithms utilize soft state, a design pattern which is known from many network protocols [8]. One possible way to implement soft state is leasing. In this case, the state of the system is only leased and has to be periodically refreshed to remain valid. If it is not refreshed in time (i.e., if it expires), it is invalidated and usually deleted. For example, we used subscription leasing to realize self-stabilizing publish/subscribe [9]. A generic implementation of self-stabilization is possible with a precautionary periodic reset [10]. In this case, all state is regularly deleted and rebuilt from an initial configuration. This ensures that corrupted state is eliminated while the correct state is established.

9 Outlook and Conclusions

We presented a modified Turing Machine which features an energy concept which is based on the laws of thermodynamics. Every program that executes correctly on this machine in the absence of errors is guaranteed to be self-stabilizing in the presence of errors. This greatly simplifies the development process since no manual proofs of the self-stabilization property is required.

They key to the self-stabilization property is that derived information has always a lower energy than the information it has been derived from. Together with the inability to amplify the energy of information we get the desired self-stabilization property. However, here we are more restrictive than the laws of thermodynamics would have required. Perhaps other machines exist which have the same self-stabilization property, but are less restrictive in some ways.

It is an open question whether the machine presented in this paper is powerful enough to execute all possible self-stabilizing algorithms. It may be the case that self-stabilizing algorithms exist which cannot execute correctly on our machine. So far we can only state the converse: if it executes correctly, then it is self-stabilizing.

Other open problems are related to the software development process. How can one easily develop applications for this kind of machine? To obtain the self-stabilization property the programs must always be executed under the control of this machine. Today's CPUs would waste much time on this. Perhaps specialized hardware could significantly improve the speed of execution.

In the future we will work on the software development process to ease the development of self-stabilizing control applications. Furthermore, a more formal description of the machine will be subject to our next publication.

References

[1] T. Weis, H. Parzyjegla, M. A. Jaeger, and G. Mühl. Self-organizing and self-stabilizing role assignment in sensor/actuator networks. In *Proceedings of DOA 2006*, volume 4276 of *Lecture Notes in Computer Science*, pages 1807–1824. Springer, 2006.
[2] M. Knoll, T. Weis, A. Ulbrich, and A. Brändle. Scripting your home. In *Proceedings of the 2nd International Workshop on Location- and Context-Awareness (LoCA 2006)*, pages 274–288, Dublin, Ireland, May 2006.

[3] Dina Q. Goldin. Persistent turing machines as a model of interactive computation. In *FoIKS '00: Proceedings of the First International Symposium on Foundations of Information and Knowledge Systems*, volume 1762 of *Lecture Notes In Computer Science (LNCS)*, pages 116–135, London, UK, 2000. Springer-Verlag.

[4] Dina Q. Goldin, Scott A. Smolka, Paul C. Attie, and Elaine L. Sonderegger. Turing machines, transition systems, and interaction. *Inf. Comput.*, 194(2):101–128, 2004.

[5] A. M. Turing. On computable numbers, with an application to the entscheidungsproblem. *Proc. London Math. Soc.*, 2(42):230–265, 1936.

[6] Edsger W. Dijkstra. Self-stabilizing systems in spite of distributed control. *Communications of the ACM*, 17(11):643–644, 1974.

[7] Shlomi Dolev. *Self-Stabilization*. MIT Press, Cambridge, MA, 2000.

[8] D. Clark. The design philosophy of the darpa internet protocols. In *SIGCOMM '88: Symposium proceedings on Communications architectures and protocols*, pages 106–114. ACM, 1988.

[9] G. Mühl, M. A. Jaeger, K. Herrmann, T. Weis, L. Fiege, and A. Ulbrich. Self-stabilizing publish/subscribe systems: Algorithms and evaluation. In *Proceedings of Euro-Par 2005*, volume 3648 of *Lecture Notes in Computer Science (LNCS)*, pages 664–674, Lisbon, Portugal, August 2005. Springer.

[10] Anish Arora and Mohamed G. Gouda. Distributed reset. *IEEE Transaction on Computers*, 43(9):1026–1038, September 1994.

Experiments with Biologically-Inspired Methods for Service Assignment in Wireless Sensor Networks

Tales Heimfarth and Peter Janacik

Abstract Given the scarcity of energy in wireless sensor networks (WSNs), in-network data processing by distributed, cooperating services is often used to reduce the amount of information that has to be routed to the base station and thereby to reduce communication and energy consumption. However, to minimize the amount of communication between services and their requesters, the locations of services in the network have to be selected carefully. Therefore, this paper proposes an efficient biologically-inspired heuristic for service assignment in WSNs. In order to reduce the amount of information exchange necessary for our heuristic, we use a concept observed in ant colonies that utilizes only local information. We model packets as ants (depositing pheromones at the visited nodes), services as food sources and requesters as formicaries. To optimize an objective function (reduction of communication distance between services and requesters), an explorer agent makes local service assignment decisions based on solely local information: the pheromones deposited by the ants. Furthermore, our paper presents the formal definition of the problem of service assignment and a thorough analysis and discussion of the results of our experiments, which show the efficiency of our approach.

1 Introduction

Wireless sensor networks (WSN) consist of a large number of embedded sensors connected via wireless links that are deployed in the monitored environment. Each node in such a network is equipped with a small processor, constrained memory, a set of sensors and, in some application examples, actuators. One key point of such

Tales Heimfarth
Federal University of Rio Grande do Sul, Brazil, e-mail: theimfarth@inf.ufrgs.br

Peter Janacik
University of Paderborn, Germany, e-mail: pjanacik@uni-paderborn.de

Please use the following format when citing this chapter:

Heimfarth, T. and Janacik, P., 2008, in IFIP International Federation for Information Processing, Volume 268; *Biologically-Inspired Collaborative Computing*; Mike Hinchey, Anastasia Pagnoni, Franz J. Rammig, Hartmut Schmeck; (Boston: Springer), pp. 71–84.

networks is the energy efficiency: since it is not feasible to replace batteries after deployment, the energy must be carefully managed in order to increase the life time of the system.

In-network data processing techniques have been successfully employed to improve considerably the energy efficiency of the network. Instead of letting each single node send its raw sensor data to the base station, which incurs a high amount of multi-hop traffic, the idea is to process the data locally in order to compute a higher level result that will be transmitted to the base station. Since nodes are only equipped with a very constrained hardware, this in-network processing, carried out by cooperating services, is distributed among neighboring nodes. Therefore, there is the need for an adequate abstraction implemented by the operating system (OS), which offers the functionality of dynamic service re-assignment to the application. This means that the OS should control the migration of the services.

We developed NanoOS [5], an OS for wireless sensor networks with the aim of supporting collaborative processing. In this work, we formalize the problem of allocating the mobile services to nodes of our network with the objective of reducing the communication overhead. Given a network topology graph and a task/service interaction graph, we aim to map the services to the nodes of the network targeting the minimization of the objective function that in our case is the communication cost. Due to the fact that our problem is NP-complete, we introduce a heuristic, responsible for the dynamic assignment of the system services within the sensor network.

This paper is organized as follows: Section 2 reviews the state-of-the-art in service assignment for WSNs, before Section 3 presents the problem definition. Section 4 introduces our ant-based service assignment heuristic, which consists of a basic and extended version. The results of the evaluation of our heuristic are then described in Section 5. Finally, Section 6 presents the conclusions.

2 Related Work

In this section, existing approaches dealing with migration of services in wireless sensor networks will be presented. Although there is a wide range of middleware and virtual machine approaches, at this moment, the majority of operating systems for WSNs do not provide service assignment mechanisms. Given the fact that most task/service assignment mechanisms used in WSNs are online (deciding during runtime), code mobility is necessary for such approaches.

In the Sensorware [3] virtual machine, the application consists of scripts deployed on a subset of network nodes. Scripts function like state machines, influenced by external events. Scripts may replicate, so that the application has the control of the service assignment, which enables agents to have individual strategies, implemented by the application programmer. In contrast, in our approach, the operating system controls the migration of services according to an OS location policy optimizing a given objective function. This disburdens the user from having to implement a migration policy in each application.

MagnetOS [1] uses the two online algorithms NetCenter and NetPull for decisions on system component assignment. NetPull monitors communication at the link level and migrates components one hop towards the neighbor with the greatest communication. NetCenter, on the other hand, relies on network-level information and migrates objects to the node hosting the component(s) they communicate the most with, possibly over multiple hops. Differently from our system, NetCenter transfers the system components directly to the node hosting the object with the highest interaction. This may lead to a non-optimal assignment and oscillations, since the sum of the communication coming from other objects at different nodes may exceed the communication traffic generated by the single component chosen as migration endpoint by MagnetOS.

In Cougar [8], queries are broadcasted to all nodes of the network and results are aggregated and forwarded to a given leader node. The query optimizer, located on the gateway node, is responsible for analyzing the queries and generating a good query execution plan, which contains the data flow inside the node and network. As this query optimizer-based approach relies on a centralized node, this and our approach are not comparable.

3 Problem Definition

In our approach we are optimizing the position of the services of the system through *migration*. Our heuristic dynamically re-assigns the services to nodes in the system in order to reduce the communication overhead. To enable the evaluation of our heuristic, we define the problem to be solved in each steady state as a formal optimization problem.

The system is represented by two graphs. The first is the network (resource) graph and the second one is the processing thread (task/service) graph (similar to the task interaction graph, TIG). The ad hoc network is modeled by an undirected graph $G = (V,E)$, where V is the set of wireless nodes and an edge $\{u,v\} \in E$ if and only if a communication link is established between nodes $u \in V$ and $v \in V$. The two nodes in this case are neighbors.

For each link, a weighting function attributes a positive weight. $w : E \rightarrow \mathbb{R}^+$. This weight measures the quality (or goodness) of a wireless link. We define for each edge not in the graph ($\{u,v\} \notin V$), $w(u,v) = \infty$. The quality of the link is calculated combining the following parameters: transmission success rate, received signal strength and history of the link. The statistic-based observation of transmission success is a good indication of the future success rate, nevertheless it reacts slowly to changes and at beginning has no data to be calculated. The received signal strength indication makes quick indications possible, but it is not very precise. Therefore, we combine these two parameters. Moreover, in order to prioritize stable links, the history is also used. We use normalized link metrics, where 0 means very good link and 1 poor one. We call the link metric *virtual distance*.

For each node, the weighting function r describes the amount of resources available at a node. $r : E \rightarrow \mathbb{R}^+$. This models the resource capacity of the node.

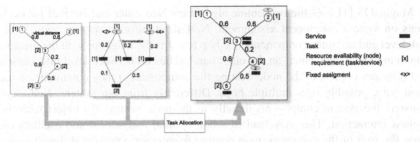

Fig. 1 Example of an instance of service assignment problem.

The processing thread (task/service) graph $T = (M,C)$ models the communication requirements between the diverse processing threads of the OS and application. M is the set of tasks and services (processing threads) running at the moment in the system and an edge $\{m_1, m_2\} \in C$ when there exist an interaction (with communication) between the executable units m_1 and m_2. For each interaction $c \in C$, a function b attributes a positive weight that measures average of traffic between the tasks/services. $b : C \to \mathbb{R}^+$. This function defines the amount of interaction between two modules of the system. Moreover, the function $e : M \to \mathbb{R}^+$ attributes the amount of resources necessary for the execution of each task/service. Finally, the function $f : M \to V$ defines the fixed assignment, i.e., the tasks that are statically assigned to a determined node and should not be moved.

The *service assignment in wireless sensor network problem* consists of allocating the tasks and services of the task graph T to the nodes of the network graph G, minimizing the amount of communication. The amount of communication is measured by the sum of all products of the amount of communication times the distance of the communicating entities. This distance is measured in terms of our link metric. A schematic diagram of the input and result of the assignment is shown in the Figure 1.

The figure 2 presents the formal definition of the optimization problem.

The problem is NP-complete (for a similar NP-complete allocation problem, see [4]), since it generalizes the well-known NP-complete quadratic assignment Problem (QAP) [7]. The QAP is a special case of our problem when the services are in the same number as the processors and just a single service (anyone) may be assigned to each processor.

4 Ant Based Service Assignment

In this section our heuristic to distribute the services in the sensor (or ad hoc) network will be presented.

Input:	A processing thread (task and service) graph with weighted nodes, weighted links, and fixed assignment function (T,b,e,f) and a network graph with weighted nodes and links (G,w,r)
Constraints:	For every input instance (G,w,r,T,b,e,f), Let $S = \{s_1,s_2,..,s_n\} = \{s \in M\|f(s) = \emptyset\}$ be the set of mobile services (without a fixed assignment). The valid solution space is given by: $$\mathscr{M}(G,w,r,T,b,e,f) =$$ $$= \{(g_1,g_2,..,g_n) \in V^n \| \forall v \in V, \textstyle\sum_{\{i \in N\|g_i=v\}} e(s_i) + \sum_{\{m \in M\|f(m)=v\}} e(m) \leq r(v)\}$$ The tuple $(g_1,g_2,..,g_n)$ is an assignment and has the following meaning: service s_i is assigned to node g_i. The constraint assures that the services and tasks assigned to the node v do not request more resource than the availability on the node.
Costs:	For every assignment $(g_1,g_2,..,g_n) \in \mathscr{M}(G,w,r,T,b,e,f)$, the cost is calculated as follows: Let the function $q : M \to V$ be: $q(m \in M) = \begin{cases} f(m) & \text{if} & f(m) \neq \emptyset \\ g_i\|s_i = m & \text{otherwise} \end{cases}$ $$cost((g_1,g_2,..,g_n),(G,w,r,T,b,e,f)) = \sum_{\{m_1,m_2\} \in C} b(\{m_1,m_2\}) \cdot D(q(m_1),q(m_2)) \quad (1)$$ Where $D(u,v)$ is the cost of the multi-hop shortest path employing the virtual distance between nodes $u,v \in V$.
Goal:	*Minimum*

Fig. 2 Formal Definition of the Optimization Problem.

4.1 Basic Heuristic

In our approach, we are optimizing the position of the services through *migration*, i.e., we try to find the optimal configuration where the communication overhead caused by the remote requests is minimized and to react to demand and topology changes adequately. In order to solve this online discrete optimization problem, we decide to use an ant-inspired algorithm. We assume, in our heuristic, that an initial distribution of the services in the network already exists. In order to describe our heuristic, some additional definitions are necessary.

The set P contains the types of all possible services of the system. Each service s is an instance of same type $p \in P$. Every task $a \in \{M - S\}$ has no type. Let $r \in M$ be the requester (a service or a task) of some service $s \in S$. The service state S_r^i represents the connection between the requester r to the service s (a flow of communication, generated by the requests and responses). The set of all flows of the system we will call W. In our system, each node $v \in V$ has a pheromone table $P_v = [p_{S_r^s}^v]_{r \in M, s \in S}$, where $p_{S_r^s}^v \in [0,1]$. This pheromone level represents the request rate (and traffic) made by the requester r to the service i that is crossing the node v. In our approach, all nodes are responsible for the service assignment, since each node's evaluation is based on its *local* view, in order to reduce communication costs. Moreover, the needed information is constantly changing, due to frequent pheromone updates.

Using an analogy with the ant foraging behavior [2], the services in our approach are the equivalent of the food source. The calls made by the requesters are the agents (or ants) and the requesters are the formicaries. The wireless links form the pathway used by the ants. While the requests are being routed to the destination service, they leave pheromone on the nodes. The pheromone tables in each node are updated according to the following equation: $p_{S_i^r}(t+1) = \frac{p_{S_i^r}(t)+\delta p(h)}{1+\delta p(h)}$ where the $\delta p(h)$ is the variation of the pheromone and it is a function of the size of the packet. After the introduction of some basic concepts of our heuristic, we will present here the component policy of our migration mechanism:

Transfer policy: In our heuristic, each service is independent and may decide it-self about starting a migration. The target of a service migration is every node with sufficient resources.

Selection policy: The selection policy is based on a threshold θ that is compared to the measured current communication overhead of the service s. If it is above θ, the service s is selected to migrate.

Location policy: The location policy decides about which node should receive a migrating service. We will describe it in the next section.

Information policy: Our heuristic uses almost just passive information gathering by means of pheromone tables. We avoid any broadcasting or proactive informa-tion dissemination to save the scarce energy resources.

The general idea is to migrate the service to some node that rely in some requests flow (path) or near to it, in the direction of a requester. Each service has several flows coming from the diverse requesters. In order to determine which node should receive the service s, an explorer packet will be used. Its next hop is defined based on the pheromone value of the neighborhood and its final location will eventually be the target node for the migration of s.

We will describe the two main phases (exploration and settlement) of the selec-tion of the new target node for the service s through the migration of the exploration packet.

Exploration Phase

In this phase, the exploration packet will migrate along the nodes of the wireless sensor network in order to find a new target position for the service s. The explo-ration phase ends and the settlement phase starts when the exploration packet has migrated a determined number of hops (allowed_h) or a loop occurred (detected using a history list *history*).

After the deployment of the exploration packet, its migration is controlled by means of attraction forces. Let $u \in V$ be the actual location (node) of the explorer packet. Ngh_u is the set of neighbors of u, and $d \in Ngh_u$ is a neighbor of u.

$$b_{u,d}^s = \begin{cases} \frac{\sum_{x \in M} p_{S_x^d}^d}{\sum_{y \in (Ngh_u-l)} \sum_{x \in M} p_{S_x^y} + pot\_pher} & \text{if} \qquad d \neq l \\ \frac{pot\_pher}{\sum_{y \in (Ngh_u-l)} \sum_{x \in M} p_{S_x^y} + pot\_pher} & \text{otherwise} \end{cases} \qquad (2)$$

$b_{u,d}^s$ represents the sum of the pheromone of all flows coming through node d to the service s normalized over the total amount of pheromone related to requests to the service s in the neighborhood. It represents relatively how much of the traf-fic directed to the service s is using the node d as path (proportional use of d for the requests). The $b_{u,d}^s$, in the exploration phase, will act as a force attracting the exploration packet to the corresponding node.

The potential pheromone (*pot_pher*) is the sum of all other pheromones related to the service s, coming from the neighbors not selected as next hop for the exploration packet *pak*, when leaving the node hosting s. It is used to estimate the level of pheromone potentially caused by those flows if the service would migrate to the node being evaluated. An example can be seen in the Figure 3.

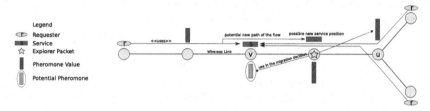

Fig. 3 Example showing the new potential path of a flow when service would migrate to the next hop.

The main idea is to predict which situation would occur if the service would migrate to the current exploration packet position and which would be the next hop for a possible migration. The assumption made here is that the request flows not attended by the first migration decision would have their path size increased exactly by the pathway executed by the exploration packet. This means, although the pheromone level from these flows would not appear to the exploration packet when far away from the node (v) hosting s, they should be considered when deciding the next exploration packet hop. This is shown in Figure 3, where the exploration packet is in the node u. It uses the real pheromone of the node j and, in the case of node v, the potential pheromone level measured by the first migration of the exploration packet. The potential pheromone level is the sum of all pheromone levels related to the service s that are in all other nodes than u because u was selected as target for the first exploration packet migration. In this example, the potential pheromone level is exactly the same level of the pheromone on node h. It will be formally defined later on.

The next hop of the explorer packet is selected using the equation 3. We call j the selected node.

$$e_i = max_{\{d \in Ngh_u\}} (b_{v \rightarrow d}^i), d \in Ngh_u \tag{3}$$

Settlement Phase

After the exploration of possible candidates to host the service s, this phase is responsible to find the appropriated node with enough resources to host the service. We call u the actual node of the exploration packet.

The idea of this phase is to evaluate whether there are enough resources at the candidate node to host the service s. In the positive case, the service will migrate to the node. In the negative one, the neighborhood will be checked and, according to the actual situation of the neighborhood, a neighbor may be selected or the exploration packet may migrate to the last visited potential candidate (retrieved from the history field), to search there for the final destination of the service s.

The following procedure is executed in the settlement phase: The current node u is tested whether it may host the service s. The test consists of checking whether node u has enough free resources. The formalization of the test can be seen in the following equation: $e(s) \leq r(u) - \sum_{\{m \in M | q(m) = u\}} e(m)$ If the resources are enough, the settlement phase is terminated and the node u sends a message to the service s to trigger the migration process. Otherwise, the same test is made in all the nodes of the direct neighborhood of u. The virtual distance is used for ordering the test process. Nodes within smallest virtual distance are tested first. The process ends when a suitable node is found, i.e., the node with enough resources and the smallest virtual distance to u is selected. We denote this node as f. If $w(u, f) < w(u, last(history))$, i.e., the virtual distance between u and f is smaller than the virtual distance between u and the last visited node by the exploration package (before reaching u), the node g is selected definitively to be the new host of s. A message is sent to s in order to start the migration. Otherwise, the exploration package is sent back to the $last(history)$ node. The node u is deleted from the history field and the settlement procedure starts again.

The procedure described above repeats until an appropriate node is found. In the rather improbable case of not finding any new node to host the service, the migration is canceled.

4.2 Extended Heuristic

This section identifies a problem caused by the greedy nature of the basic heuristic and presents an improvement to overcome possible adversarial situations. For the sake of simplicity, we assume in the following example that allowed_h=1, i.e., just one hop migrations are allowed. Nevertheless, the problem occurs for arbitrary values of this parameter when more than one nearby located requesters use the same service, but due to the employed routing algorithm, the requests are routed through different paths. An example of such situation is depicted in Fig. 4, where requesters r_1, r_2 and r_3 are accessing the service s in the node u. For a straightforward communication cost calculation, we assume that the average bandwidth utilization is proportional to the pheromone deposited at a node inside the flow path. Thus, the total communication cost is 1.135 (using equation 1).

We analyze the migration that would be decided by the basic heuristic. As the pheromone value of node h is higher than the values deposited at nodes j and k (separately), the exploration packet is sent to node h. Suppose that allowed_h=1, the service would migrate to node h. The total communication cost of the system changes to 1.22. This result shows that the heuristic, in such adverse situation, selects the wrong node to migrate to, increasing the total communication cost of the

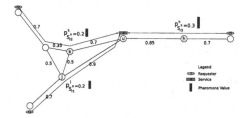

Fig. 4 Instance of the problem that will result in a wrong migration decision due to greedy behavior.

system. This happens because of the lack of information over not directly-connected parts of the network (each node has just the *local* view of the system).

The main idea of the improvement is not to migrate the service to the neighbor with the highest amount of requests (highest flow) as in the basic heuristic, but to the neighbor whose flow, in some part, is crossing nodes near to other flows requesting the same service. If the defined metric (virtual distance) has (geographical) norm properties, this will be equivalent to migrating the service to the geographical *direction* from where the highest amount of requests is coming. Two flows related to the requesters r_1 and r_2 (see Fig. 4) are transversing neighboring nodes in their path to s, thus, they should attract the service instead of r_3. We define that such flows transversing neighboring nodes are called correlated flows.

The concept of the correlated flows is used in the exploration phase in order to guide the migration of the exploration packet. Instead of counting solely the pheromone deposited at each neighbor when analyzing the amount of pheromone of a neighbor, the sum of the pheromone deposited at the node with all correlated pheromone is used to guide the migration. Therefore the equation 2 is modified as follows:

$$
b^s_{u,d} = \frac{\overbrace{\sum_{x \in M} p^d_{S^s_x}}^{\text{flows using }d} + \overbrace{\sum_{x \in M} \sum_{z \in M} \sum_{g \in Ngh_v - \{d,l\}} p^g_{S^s_z} \cdot \lceil p^d_{S^s_x} \rceil \cdot F(S^s_z, S^s_x)}^{\text{correlated flows}}}{normalizer} \quad (\text{if } d \neq l) \quad (4)
$$

The first term of the equation is the same as in eq. 2, i.e., the sum of all requests coming to service s through node d. The second term of the numerator is the sum of the pheromone generated by correlated flows of the flows present at node d. The function F tests whether S^s_z and S^s_x are correlated flows, and the ceiling $\lceil p^d_{S^s_x} \rceil$ checks whether the connection S^s_x exists in the node d (i.e. $p^d_{S^s_x} > 0$). The denominator normalizes $b^s_{u,d}$ ($0 \leq b^s_{u,d} \leq 1$). The next hop of the exploration packet is selected using equation 3.

5 Results

In this section, we present the simulation results of our basic and extended service assignment heuristics. The simulations were performed using the Shox [6] simulator, an event-based wireless network simulator. For our simulation, we assume fixed transmission power, bidirectional links (which is achieved in reality by ignoring unidirectional links) and Friis Free Space propagation model for isotropic point source in an ideal propagation medium for RSSI calculations. The link metric used is based on the RSSI and each node only offers enough resources for running a single service and task. Tasks request different, randomly selected services. The bandwidths needed in the different communications were randomly selected.

5.1 Simulation Scenarios and Evaluation

Table 1 provides an overview of the simulated scenarios. The small scenarios were selected since they also allow the calculation of the optimal solution. For large scenarios, it is not possible to calculate the optimal (reference) solution of our discrete optimization problem due to its computational complexity. Nevertheless, we decided to make an example simulation of a large scenario to show that its behavior is similar to small instances.

For the generation of the task/service graph for each task, a random number of services was selected. The tasks request those services with a random bandwidth requirement (normalized). Dijkstra's shortest path algorithm was used for finding routes between requesters and the services.

Table 1 Overview of the different simulation scenarios.

Scenario Name	Field Size (m^2)	Number of Nodes	Radio Range	Connection Probability	Node density	Average Degree (Theoretic)	Num. Services, Requesters
Small Scenarios							
small-sparse-sd	80x60	10	28	0.9	0.002	5.13	8, 6
small-dense-sd	80x60	10	43	1	0.002	12.1	8, 6
Large Scenarios							
large-sparse-sd	102x77	100	13	0.9	0.013	6.7	20, 40

The presented scenarios were evaluated using different algorithms. For the small scenario, the optimum solution was calculated using a branch-and-bound algorithm. For all scenarios, our basic and extended ant-based service assignment heuristics were simulated. Moreover, we decided to calculate the cost of a completely random assignment, i.e., the tasks and services are randomly distributed among the nodes of the network.

5.2 Experiments

We executed 40 experiments for each scenario presented in Table 1. In the next sections, we will present and analyze the results of the experiments.

Optimal Assignment Cost

In this section, we will analyze the results achieved with the optimal cost assignment. Figure 5(a) presents the optimal service assignment cost for the small-sparse-sd and small-large-sd scenarios[1]. As expected, denser scenarios exhibit a smaller assignment cost. This can be explained by the fact that better links (lower cost) are available for the communication between tasks and services, reducing the total cost. Moreover, due to the higher amount of neighbors (higher node degree), assignments that yield a high amount of costs in sparse environments may be attractive in dense environments because of the existence of multiple new links.

Experiment Results

This section presents the outcome of our 40 experiments for the presented scenarios. In Figure 5, the results for our three scenarios are depicted. For the small scenarios, each result is normalized against the optimal assignment. For the large scenario, we present the nominal result.

As we can see in Figures 5(b) and 5(c), our heuristic found the optimal solution in several cases. Moreover, for the vast majority of cases, the heuristic has a much better performance than the random initial assignment. The extended heuristic and the basic one have also a very similar behavior, nevertheless, for some experiments, the extended one has a much better performance than the basic. The reasons for these outcomes will be discussed further below.

Figure 5(d) shows the results for a large scenario. Due to the fact that we do not have, for large scenarios, a reference approach, it is not possible to make statements about the absolute performance of the algorithms. Nevertheless, it is possible to notice that the heuristics could find a much better cost than the initial random assignment. Moreover, the behavior of the extended and basic heuristics are similar to the one observed in the small experiments.

Heuristics' Assignment Costs

In this section, the mean value of the achieved costs for each heuristic for all scenarios will be presented. The cost for the absolute assignment of our test scenario is shown in Figure 6. The random assignments and basic and extended heuristic assignments have the same tendency of the optimum: for sparse networks, they deliver always an assignment with higher cost. This is expected due to the relation between the assignment cost and the link costs, and for sparse environments, the average link cost increases.

In Figure 6(a), the different costs are shown together for the small scenarios. As it can be seen in the figure, our basic and extended heuristics have a good performance,

---

[1] For the service assignments, the terms total communication cost (presented in the figure) and assignment cost have the same meaning.

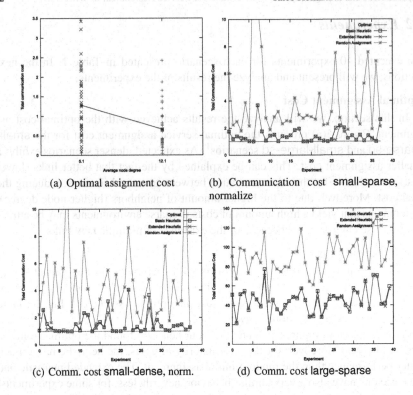

(a) Optimal assignment cost   (b) Communication cost small-sparse, normalize

(c) Comm. cost small-dense, norm.   (d) Comm. cost large-sparse

Fig. 5 (a) Optimal assignment cost of sparse and dense scenarios and (b–d) communication cost results of the realized experiments.

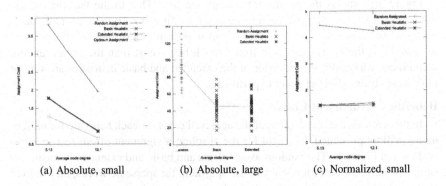

(a) Absolute, small   (b) Absolute, large   (c) Normalized, small

Fig. 6 Assignment costs for the different heuristics with small and large problem size.

not far from the optimal solution. The basic and extended heuristics have a very similar performance. We discuss the reasons and the performance difference further below.

In Figure 6(b), the results of our large scenario are depicted. It is possible to see that they are very similar to the small scenario, improving our confidence that the heuristics could find good solutions for small as well as for large scenarios.

Figure 6(c) shows the normalized results for the small scenarios. The optimal assignment is used as reference. It is possible to notice that, for all cases, a very small difference could be verified for sparse and dense scenarios. The basic heuristic has an average cost of 1.44 times the optimal cost for sparse environments and 1.5 for dense ones. The extended heuristic shows a small improvement: 1.41 for sparse and 1.43 for dense scenarios. This means that the cost of the basic heuristic was about 2% higher in sparse scenarios and 5% in dense scenarios. A similar behavior has been found in the large scenario.

As it can be observed in Figures 5(b) and 5(c), the basic and extended heuristic, for several experiments, could find solutions with very similar costs and for some experiments, the extended overcame the basic one. For the experiments where the results were similar, we suppose that there are not flow correlations that help the heuristic behavior. For the experiments where the extended heuristic has a much better performance, correlations could be found and a better service migration was realized.

The question that arises from the results is why correlations were not so common. We suppose that the reason was the selected routing algorithm together with the influence of the Friis Free Space Model in our link metric. Because of the exponential path loss, nodes near to each other have a greater advantage in the signal strength than others with a small higher physical distance. Due to the fact that we are, for our simulations, relying strongly on the signal strength to calculate the link metric, it reflects very much this exponential path loss. The Dijkstra's shortest path algorithm always selects the shortest path between any two nodes and does not try to divide the load among the existing link channels. Further we are also not taking into account the link utilization (and possible congestion). Together, such facts act in a way that effectively just a small subset of links is used for all communications. A kind of backbone emerges in the network. This leaves less space for our flow correlations. We suppose that, in real scenarios, where the link metric has a more irregular nature and where there are routing mechanisms that divide the transmission effort among different routes, the extended heuristic will increase its performance in relation to the basic one.

6 Conclusion

In this paper, we study the problem of automatic assignment of mobile services on wireless sensor networks. We model our problem as an optimization problem and present an efficient biologically-inspired heuristic to solve it. Given an initial assignment, the heuristic is responsible to drive the migration of the services based on the actual network/service configuration targeting the reduction of the communication cost.

We chose to use concepts observed in ant colonies, since they exhibit several properties that are desirable in wireless sensor networks. Hence, we propose an extension to the heuristic: neighboring pheromone trails act together to attract the service to the direction with higher request rate.

Simulations done using the wireless network simulator Shox showed that the heuristic perform well. The basic heuristic has an average cost of 1.44 times the optimal cost for sparse environments and 1.5 for dense ones. The extended heuristic produces a slighter better result than the basic one: 1.41 for sparse and 1.43 for dense scenarios. We suppose that this occurs due to the fact that just a small subset of links is used to route almost all packets in the network (a backbone is formed).

For a large number of real applications, the basic heuristic yields adequate results with very small computational cost. The extra effort necessary for the extended heuristic may not be compensated. However, in real environments, where the link metric does not follow in a regular way the Friis path loss and different routing mechanisms may be used, the extended heuristic may bring better results.

Concluding, our work provides an additional piece of evidence that concepts inspired by biology can be successfully transferred to computer systems.

References

1. Rimon Barr, John C. Bicket, Daniel S. Dantas, Bowei Du, T. W. Danny Kim, Bing Zhou, and Emin Sirer. On the need for system-level support for ad hoc and sensor networks. *SIGOPS Oper. Syst. Rev.*, 36(2):1–5, 2002.
2. Eric Bonabeau, Marco Dorigo, and Guy Theraulaz. *Swarm Intelligence: From Natural to Artificial Systems*. Oxford University Press, New York, NY, 1999.
3. A. Boulis and M. B. Srivastava. Design and implementation of a framework for efficient and programmable sensor networks. In *Proc. of the First International Conference on Mobile Systems, Applications, and Services (MobiSys 2003), San Francisco, CA, USA*, San Francisco, CA, USA, May 2003.
4. David Fernandez-Baca. Allocating modules to processors in a distributed system. *IEEE Transactions on Software Engineering*, 15(11):1427–1436, November 1989.
5. Tales Heimfarth and Achim Rettberg. Nanoos - reconfigurable os for embedded mobile devices. In *In Proceedings of the International Workshop on Dependable Embedded Systems (WDES)*, Florianopolis, Brazil, 2004.
6. Johannes Lessmann, Tales Heimfarth, and Peter Janacik. Shox: An easy to use simulation platform for wireless networks. In *Proceedings of the 10th International Conference on Computer Modeling and Simulation*, Cambridge, England, Apr. 2008.
7. Sartaj Sahni and Teofilo Gonzalez. P-complete approximation problems. *J. ACM*, 23(3):555–565, 1976.
8. Yong Yao and Johannes Gehrke. The cougar approach to in-network query processing in sensor networks. *SIGMOD*, 31(3), September 2002.

Evolving Collision Avoidance on Autonomous Robots

Lukas König and Hartmut Schmeck

Abstract Utilizing the collective behavior of a population of interacting individuals, based on rather simple local algorithms, is a promising approach for achieving complex goals. We use an onboard online evolutionary model, based on finite Moore automata, to develop collective behavior in an artificial swarm of micro-robots. Experiments have been made in simulation to achieve Collision Avoidance. The model is shown to be capable to generate the desired behavior and we present experiments for adjusting the parameters of the evolutionary optimization.

1 Introduction

As it has been shown in the past [1, 2], collective behavior can reduce the algorithmic complexity of solving tasks, by exploiting emergent effects in a swarm originating from the interaction between its single individuals. Modern robotics can profit from this concept. However, the emergent collective behavior is hard to predict, and given a task, it is generally not obvious, how programs should be designed to provide, as a collective behavior, the solution of this task. In some cases, one can develop algorithms and strategies manually, but in general, this turns out to be hard even for simple tasks and non-collective applications.

A common strategy for finding potential templates or strategies for the design of a collective behavior is to observe swarms in nature. They show adaptive behavior and, quite often, they are at least close to solving some of their environmental challenges in an optimal way, the most prominent, standard example being the capability of ant colonies to find shortest paths [3]. By a careful analysis of the behavior of natural and artificial swarms, one can hopefully extract appropriate local rules for collective behavior [1]. However, there is no guarantee that the behavior of a natural swarm can be understood sufficiently well to successfully generate the required behavior for an artificial swarm. Also, there are many conceivable tasks for artificial swarms which do not have a related counterpart in nature.

Lukas König
University of Karlsruhe, AIFB, e-mail: koenig@aifb.uni-karlsruhe.de

Hartmut Schmeck
University of Karlsruhe, AIFB, e-mail: schmeck@aifb.uni-karlsruhe.de

Please use the following format when citing this chapter:

König, L. and Schmeck, H., 2008, in IFIP International Federation for Information Processing, Volume 268; *Biologically-Inspired Collaborative Computing*; Mike Hinchey, Anastasia Pagnoni, Franz J. Rammig, Hartmut Schmeck; (Boston: Springer), pp. 85–94.

One of the suggested approaches to generate local tasks for collections of cooperating agents is to use evolutionary algorithms or genetic programming [4]. In this paper, we use an *onboard online* evolutionary approach to develop collective behavior in a population of robots. *Onboard* means that each robot has an evolutionary program running, which is separated from the other robots and especially is not triggered by some kind of central control with global information. *Online* means that during a run, the robots are supposed not only to achieve the ability to solve a given task, but also to solve the task in the given environment in order to evaluate the feasibility of the current solution. So the idea is to confront collections of robots with a problem, and to let them learn cooperatively through evolution, until an adequate solution is found, and to solve the problem at the same time.

Based on the approach presented in [5] the evolutionary model is built on finite Moore automata and it is defined in a general way to be applicable on different robot platforms. Up to now it has been implemented and tested on the Jasmine IIIp robot platform at the University of Stuttgart and in simulation, where also the Jasmine IIIp robot has been modeled. The Jasmine IIIp series is a swarm of micro-robots sized $26 \times 26 \times 26mm^3$. It can drive forwards and backwards and turn left and right. Each robot has seven infra-red sensors (two facing to the front, the others being placed in steps of 60 degrees around, each returning values between 0 and 255) to measure distances to obstacles and to communicate with other robots (cf. www.swarmrobot.org).

This paper extends the approach presented in [5] by using extensive simulation experiments to adjust evolutionary parameters and to show that Collision Avoidance can be evolved.

In Sec. 2, we describe the theoretical model and the implementation of the framework for the Jasmine III robot in simulation. In Sec. 3, results of the evolutionary runs are presented. Sec. 4 provides a conclusion and an outlook to future work.

2 Model Description

The developed behavioral model for robots is based on finite Moore automata defined in a common manner [6]. The output of a state defines an instruction to be executed. The transitions depend on the internal state and on the information provided by the sensors. The automaton is referred to as the *genome* of the robot, while the resulting behavior (i.e., the mapping from a sequence of sensor data to the corresponding sequence of output instructions) is called the phenotype; accordingly, the *genotypic search space* Γ is defined to be the space of all Moore automata, while the *phenotypic search space* Ω is the space of all behaviors. The genome can be modified by mutation and crossover.

Due to space limitations, this section only provides a brief overview of the model and the implementation. For more detailed information see [5].

Preliminaries. We denote a set of byte values and a set of positive byte values as $B = \{0,...,255\}$ and $B_+ = \{1,...,255\}$. The behavior depends on sensor data represented by a set H of n sensor variables $H = \{h_1,...,h_n\}$. The sensor data may originate from real or virtual sensors (the latter being any internal variables of the robot). Each variable h_i can be set to any byte value. The seven main infra-red sensors of the robot are stored as $h_1,...,h_7$, starting with the two sensors facing to the front and then incrementing clockwise. We did not use any other sensors for the experiments.

We assume a set $I = \{I_1,...,I_m\} \subseteq B_+$ of m instructions, encoded as positive byte values. In general, instructions may be interpreted as arbitrary programs, which are capable to run on a robot; however, up to now only the following simple instructions have been used: (1) Idle (i. e., "keep executing the last instruction"), (2) Stop, (3) Move forward, (4) Turn left, (5) Turn right.

We assume a function *rand*, which returns a random element out of an arbitrary finite set, based on uniform distribution.

Moore automaton for robot behavior. A finite Moore automaton for robot behavior (MARB) is a Moore Automaton $A = (Q, \Sigma, \Omega, \delta, \lambda, q_0)$, where:

Q is the set of states (q_0 being the initial state).

The input alphabet $\Sigma = B^n$ consists of all possible combinations of sensor values.

The output alphabet $\Omega = I \times B_+$ consists of the instructions with an additional parameter.

The transition function δ is defined for any state and each combination of sensor values in Σ by specifying for each state q a list $((c_1, q_1), (c_2, q_2), ...)$ of conditions and associated following states; it is interpreted like a case-statement, i. e., the first condition evaluating to *true* under the current input determines the next state. The conditions are conjunctive and disjunctive combinations of *false*, *true*, or relational expressions of the type $a\ rel\ b$, where $a, b \in H \cup B_+, rel \in \{<, >, \leq, \geq, =, \neq, \approx, \not\approx\}$. $\approx$ is true ($\not\approx$ is false) whenever the two operands differ by at most a constant (which is set to 5 in our experiments). If none of the conditions evaluates to true there is a default transition to the initial state (see Fig. 1).

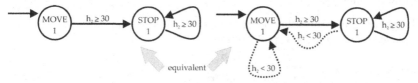

Fig. 1 Example of implicit transitions to the initial state if no condition is *true*.

The output function λ assigns to each state an instruction with a parameter, e. g., (Turn left, 45), which lets the robot turn left by 45 degrees. The parameter can be any positive byte value.

Mutation. The mutation operator is defined as a mapping in the genotypic search space: $M : \Gamma \rightarrow \Gamma$ (i. e. it maps a MARB A into a MARB $M(A)$). Let $k \in \{0,...,255\}$ be a constant (we set $k = 5$ in all experiments). A mutation randomly selects one of the following atomic transformations:

1. Toggle "inactive" transitions (*syntactic*, i. e., no change of behavior): Remove a random transition, associated with the condition *false* or add a random transition, associated with the condition *false*.
2. Remove an "inactive" state (*syntactic*): Remove a state q without incoming transitions or with all outgoing transitions being associated with the condition *false* and the state being associated with the instruction *IDLE*.
3. Add a new state q (*syntactic*): q has no incoming transitions and no outgoing transitions, random instruction and a random parameter $\leq k$.
4. Change a condition: Let $a,b \in B_+ \cup H$, c a condition. Any part of a condition that matches the following patterns can be mutated (the notation $x \leftrightarrow x'$ means that x may be replaced by x' and vice versa.):

   a. (*semantic*, i. e., potential change of behavior)
   $$false \leftrightarrow a = b \leftrightarrow a \approx b \leftrightarrow \begin{matrix} a \leq b \leftrightarrow a < b \\ a \geq b \leftrightarrow a > b \end{matrix} \leftrightarrow a \not\approx b \leftrightarrow a \neq b \leftrightarrow true$$
   b. One of the following (*syntactic*):

(c AND $true$)	$\leftrightarrow$	c	$\leftrightarrow$	($true$ AND c)
(c OR $true$)	$\rightarrow$	$true$	$\leftarrow$	($true$ OR c)
(c AND $false$)	$\rightarrow$	$false$	$\leftarrow$	($false$ AND c)
(c OR $false$)	$\leftrightarrow$	c	$\leftrightarrow$	($false$ OR c)

   c. Change a number i within a condition (*semantic*): Let $i' = i + rand(\{-k,...,k\})$, Replace i as follows:
   $$i \rightarrow \begin{cases} i', & \text{if } 1 \leq i' \leq 255 \\ 1, & \text{if } i' < 1 \\ 255, & \text{if } i' > 255 \end{cases}.$$

   d. Change a sensor variable h within a condition (*semantic*): Replace h with $rand(H)$.

5. Change a state q (*semantic*): Let (I,P) be the output of q. Replace (I,P) with $(J, |P + c| + 1)$, where
   $$c = rand(\{-k,...,k\}), J = \begin{cases} I, & \text{if } P + c > 1 \\ rand(I) & \text{otherwise} \end{cases}.$$

Mutation was performed once within a time interval S. This interval was studied in the experiments (see Sec. 3).

Obviously, an appropriate sequence of mutations can transform any MARB A into any other MARB A' by changing its topology, conditions, and the output function (i. e. the mutation operator is complete). In order to make the mutations "smooth" in the sense that a single mutation causes only a "small" change in the phenotypic search space, the focus is on keeping the semantic mutations (i. e., mutations that potentially change the behavior) few and small.

Reproduction/Selection. Since in the onboard concept there is no central unit with information about the whole population, it is not possible to implement a global selection operator. Instead, similar to the diffusion model of evolutionary algorithms (cf. [7]) we use local selection: A robot produces a child genome together with its closest neighbor. As in [5], a very simplified recombination operator is used which assigns to both robots the parental genome having the better fitness. Note, however, that this could easily be replaced with a more standard stochastic crossover operator combining parts of both parental genomes into a child genome. Future work will feature such a crossover operator.

On real robots, reproduction is performed each time two robots meet, i. e., come closer to each other than some threshold. In simulation, we implemented a similar solution which, however, is easier to analyze: Using a constant time interval T, each robot reproduces with the robot which, after T time units, is the spatially closest to itself. However, this does not mean that all robots recombine simultaneously; for each robot a separate timer is running which, due to possibly delayed requests, drifts apart during the experiment. The parameter T was studied in the experiments (see Sec. 3).

Fitness function. As the fitness of a MARB has to be evaluated locally, it has to be based on the observed sequence of sensor data which is influenced by the generated behavior of the robot. Therefore, every U time units, the fitness value is updated by a "fitness snapshot" (see below).

Since mutations modify the behavior, the fitness value has to be adjusted. This is done by using "evaporation", i. e., every V time units, the fitness value of the robot is divided by 2. U and V are parameters that are studied in Sec. 3. Furthermore, undesirable events (like collisions) should modify the fitness appropriately.

For *Collision Avoidance*, we used a fitness measure, which states that moving around is good, but being near an obstacle is bad; colliding with an obstacle is even worse. The fitness assignment is shown in Alg. 1. It holds that NOT_MOVING is 0, if the robot's current instruction is "Move", 1 otherwise; $OBST\_NEAR$ is 1, if a close obstacle is sensed (i. e., $\exists h \in H : h > 100$), 0 otherwise. Initially, the fitness value of every robot is set to 0.

```
if *U expired* then   // Add snapshot to fitness.
  fitness += (1 - NOT_MOVING - OBST_NEAR);
end if
if *Collision* then fitness -= 3; end if
if *V expired* then fitness /= 2; end if
```

Alg. 1: Fitness assignment and update.

3 Experiments

When doing evolutionary computation, it is usually required to adjust a set of parameters, before good results can be achieved. We made experiments to adjust the four parameters mutation interval S, reproduction interval T, fitness snapshot interval U and fitness evaporation interval V. However, there are more parameters than these four, which have to be studied in future experiments (e. g., number of robots, size of field, and more complex parameters like the mutation and crossover operators).

Collision Avoidance as target behavior has been used, because it is a simple and analyzable behavior. However, as it was not a priori clear if this is even evolvable with a model based on finite Moore automata and how it would work, we present also two of the resulting automata.

Since there was a large number of evolved robots (in total 21 060 robots in 810 simulations), it was not possible to look at all results in detail. Instead, we checked only those automata, which finally achieved a positive fitness value. To justify this, we separately tested automata with fitness > 0 (A) and those with fitness ≤ 0 (B), whether they had a reachable "Move"-state (since otherwise, they could not move and, therefore, especially did not perform Collision Avoidance). It turned out that 91% of the automata in (A) had a reachable "Move"-state, but only 28% of those in (B) did. Also, the observation of random samples convinced us that zero is a reasonable fitness threshold to distinguish between "good" and "bad" Collision Avoidance behavior.

Adjusting the parameters. As listed in Tab. 1, $3^4 = 81$ different parameter combinations have been used, setting each of the four parameters to three constant values. Each of these experiments was repeated 10 times. The setting was a rectangular field, $60 \times 80cm$, where 26 robots were placed randomly (based on uniform distribution in position and angle). Their initial automaton was completely empty, i. e., had no states. Simulation was performed for about 2000000 simulation steps (a robot would drive about 8 km in one experiment if only repeating the Move-instruction).

Fig. 2 shows the number of "successful" experiments, i. e., experiments in which there existed robots with a positive fitness in the end, distributed on the 81 different parameter combinations. On average, there were about

Table 1 Parameter values used in the experiments.

	Mutation Int. S	Reproduction Int. T	Snapshot Int. U	Evaporation Int. V
1.	5000 ms	1000 ms	250 ms	10000 ms
2.	10000 ms	2000 ms	500 ms	20000 ms
3.	20000 ms	10000 ms	1000 ms	30000 ms

7 robots with a positive fitness in the final populations of successful experiments (see Sec. 4 for a discussion on selective pressure).

As Tab. 2 shows, some of the results indicate a tendency, in which direction parameters should be shifted.

Table 2 Distribution of successful experiments for each parameter separately.

	Mutation Int. S	Reproduction Int. T	Snapshot Int. U	Evaporation Int. V
1.	5000 ms: 18%	1000 ms: 37%	250 ms: 37%	10000 ms: 31%
2.	10000 ms: 29%	2000 ms: 29%	500 ms: 36%	20000 ms: 30%
3.	20000 ms: 53%	10000 ms: 34%	1000 ms: 27%	30000 ms: 38%

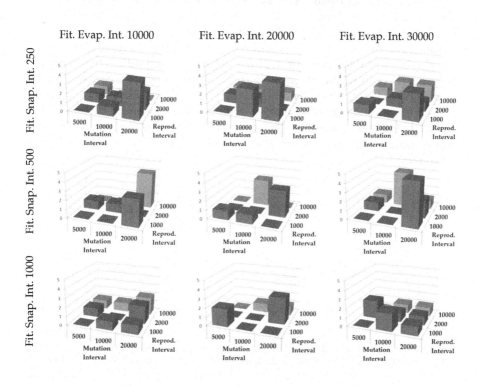

Fig. 2 Distribution of successful experiments (i. e., number of experiments out of 10 repetitions, where at least one robot had a positive fitness – for the 81 parameter combinations).

The number of successful experiments increases with a larger *mutation interval S*. In particular, a mutation interval of 5000 *ms* almost never yielded successful experiments (18%). So, looking at all parameter combinations, it seems to be reasonable to increase the mutation interval even more.

For the *reproduction interval T*, no clear statements can be derived. Apparently, the values between 1000 *ms* and 10000 *ms* have to be studied more precisely.

For the *fitness snapshot interval U*, it seems as if the value should be decreased. However, the differences are smaller than at parameter *S*.

For the *fitness evaporation interval V*, higher values seem to be better.

However, due to dependencies between the parameters, it may not be possible to find the perfect parameter combination by only optimizing each parameter separately. Rather than doing that, we will continue to perform experiments with a large set of parameter combinations to learn more about these dependencies. We will use these results, however, to draw conclusions about the directions, in which the search should be extended.

The evolution of Collision Avoidance. In our experiments, 545 robots (2.6%) had a positive final fitness. Some of these achieved the expected Collision Avoidance behavior, i. e., moving (arbitrarily) until an obstacle appears, then turning until the way is clear, then moving again; some did some other forms like driving in circles or ellipses. However, it is often hard to characterize a behavior accurately, since it could depend on circumstances, which are hard to understand; e. g., there were robots which avoided obstacles only when a specific constellation of sensor values from the sensors in the back was received. Therefore, we characterize a robot's behavior only by its fitness value; in future the fitness function should be refined to avoid unwanted behaviors.

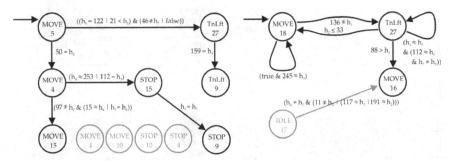

Fig. 3 Two examples of evolved automata – performing the expected Collision Avoidance behavior (left, fitness: 730); driving a circle (right, fitness: 132); unreachable states set gray.

Fig. 3 shows two automata that evolved during the experiments. The left one lets the robot essentially drive straight forward, except when sensing an obstacle: then it turns left. The right one is driving a circle – without sensing obstacles at all. However, due to the implementation of crash simulation, it turns out that a circle-driving robot finds, after a few initial collisions, enough space to drive with nearly no colliding and, therefore, is successful in the sense of the fitness function.

4 Discussion

Conclusion. The results presented in this paper show that the onboard approach of online evolution of robot behavior based on a Moore Automaton control works in principle. A set of four parameters has been selected quite arbitrarily and set to some almost arbitrary values to study their influence on the quality of the resulting behavior with respect to the fitness function. Even within this random selection of the parameter search space, a parameter combination has been found, which in 5 out of 10 samples yielded successful experiments (and others did so in 3 or 4 out of 10 trials). It can be expected that even better parameters can be found by extending the search in the directions indicated by the results of the experiments and by modifying other parameters, which have not been considered in this paper.

However, the experiments also uncovered some problems, which have to be considered in future work. The process of reproduction, which is also the mechanism of selection, is based on the assumption that the robots are moving around at least now and then. If too many are standing idle, the rule of producing offspring always with the closest neighbor determines an incestuous exchange of genomes, always with the same partner. Since we argued that most of the robots had a negative fitness in the end and that of their automata only about 28% even had a reachable Move-state, we can expect that this assumption was not fulfilled sufficiently in many populations. Since a similar problem arose in experiments with real robots, the movement of robots during evolution has to be studied more carefully and probably a new mechanism of reproduction has to be developed. Maybe using a flexile mutation interval (e.g., mutating robots with lower fitness more often), rather than a constant one as we did so far, could also help avoiding this. A second problem was the fitness function, which induced unexpected solutions like driving circles without sensing obstacles. Though these solutions were successful in the sense of the fitness function, it was not the intended behavior. Obviously, the design of a fitness function for a target behavior has to be studied in correlation with the expected and the actual outcome.

Another issue is the measure of diversity (as defined in [8]) in the population during the experiments, which could give a more detailed insight into the evolutionary process. Besides inherent problems of the diversity measure, in this case, it is not even clear, how to measure the difference between two single individuals. As proposed in [9], this difference could be measured on the phenotypic level through the reaction of individuals to stimuli (i. e., some kind of fitness function, again). However, the fitness function is delayed and dependent on the environment, so a more precise measure on the genotypic level should be found. For two automata A and A' such a measure could, e. g., be based on the number of input sequences (i. e., sequences of sensor combinations), where the corresponding output sequences produced by A and A' differ.

So far, we considered fitness values and genomes at the end of experiments, only. However, fitness, genome, and movement of each robot are recorded during the experiments. A careful analysis of this data should allow to gain more information about selective pressure, the benefits of a flexible mutation interval, the problem with incestuous exchange of genomes and it could help to get a better idea of an appropriate evolving time.

Outlook to future work. More experiments with a larger variety of parameter combinations have to be performed and the statistical significance of the results has to be checked. Especially studying the mutation interval seems to be promising for achieving better results. A recombination operator which generates offspring as a mixture of the genomes of both parents has already been developed, but still has to be implemented and tested. Also, we are planning to make experiments with a flexible mutation interval. A measure for diversity on the genotypic level is being developed. The development of the population during evolutionary runs is going to be studied in detail.

References

1. Bonabeau, E.; Theraulaz, G.; Dorigo, M.: Swarm Intelligence. From Natural to Artificial Systems. Santa Fe Institute Studies in the Sciences of Complexity, Oxford University Press, 1999
2. Kube, C. R.; Zhang, H.: Collective Robotics: From Social Insects to Robots. Adaptive Behavior, 2, 2, 189-218, 1993
3. Goss, S.; Aron, S.; Deneubourg, J. L.; Pasteels, J. M.: Self-organized shortcuts in the argentine ant. Naturwissenschaften, 76, 579-581, 1989
4. Branke, J.; Schmeck, H.: Evolutionary design of emergent behavior. In "Organic Computing", Series: Understanding Complex Systems. Wrtz, Rolf P. (Ed.), Springer Verlag, 123-140, 2008
5. König, L.; Jebens, K.; Kernbach, S.; Levi P.: Stability of on-line and on-board evolving of adaptive collective behavior. Euros, Prague (accepted), 2008
6. Hopcroft, J. E.; Ullman, J. D.: Introduction to Automata Theory, Languages, and Computation. Addison-Wesley, 1979
7. Schmeck, H.; Branke, J.; Kohlmorgen, U.: Parallel implementations of evolutionary algorithms. In: Solutions to Parallel and Distributed Computing Problems, Zomaya, A.; Ercal, F.; Olariu, S. (Ed.). John Wiley, New York, Wiley Series on Parallel an Distributed Computing, 47-66, 2001
8. Nehring, K.; Puppe, C.: A Theory of Diversity, *Econometrica*, 70, 1155-1198, 2002
9. Curran, D.; O'Riordan, C.: Increasing population diversity through cultural learning, *Adaptive Behavior*, 14, 4, 2006

Local Strategies for Connecting Stations by Small Robotic Networks*

Friedhelm Meyer auf der Heide and Barbara Schneider

Abstract Consider a group of m stations with fixed positions in the plane and a group of n mobile robots, called relays, aiming at building a communication network between the stations consisting of as few relays as possible. We present two strategies for dimensionless, identical (anonymous), oblivious and disoriented relays with limited viewing radius for constructing such a network. These strategies resemble natural strategies of swarms for maintaining formations. A relay does not communicate with others, its decision - whether to remove itself from the system, or where to move - consists only of the relative positions of its neighbors within its viewing radius. We provide a theoretical analysis of worst-case scenarios and upper and lower bounds for the number of relays used by the strategies. In addition, we show some preliminary experimental results.

Keywords: Mobile robots; self-organization; distributed computing; cooperation; ad-hoc networks.

1 Introduction

In our research we investigate the construction of a communication network connecting stations in a planar terrain without obstacles. Mobile relays are used to route messages between the stations. Both stations and relays have a restricted viewing

Friedhelm Meyer auf der Heide
Heinz Nixdorf Institute and Computer Science Department, University of Paderborn e-mail: fmadh@upb.de

Barbara Schneider
Computer Science Department, University of Paderborn e-mail: barbaras@upb.de

 * Partially supported by the European project "Foundations of adaptive networked societies of tiny artifacts (FRONTS)" within the 7th Framework programme, and the DFG-project "Smart Teams" within the SPP 1183 "Organic Computing"

Please use the following format when citing this chapter:

Heide, F.M.a.d. and Schneider, B., 2008, in IFIP International Federation for Information Processing, Volume 268; *Biologically-Inspired Collaborative Computing*; Mike Hinchey, Anastasia Pagnoni, Franz J. Rammig, Hartmut Schmeck; (Boston: Springer), pp. 95–104.

radius and thus are connected via their so-called unit disk graph. They are able to sense the positions of all nodes within this distance. Starting with a large group of relays with positions somewhere in the plane and a group of stations with fixed positions, our goal is to develop local strategies for the movement and removal of the relays that globally result in a network of relays connecting the stations using as few relays as possible. We assume that the initial unit disk graph of the stations and relays is connected and our strategies must keep it connected. A globally optimal solution would place the relays on the edges of a Minimum Steiner Tree connecting the stations.

In Section 2 we present two local, distributed strategies for the movement and removal of the relays. Both versions are very intuitive and do not use any communication, the relays perform their movements using only the sensed positions of the relays and stations within the communication distance. This approach is defined in [1] and called "Interaction via sensing". In Section 3 and 4 we analyze and compare the performance of the strategies concerning the number of relays used. The theoretical results provide a worst-case analysis, the experimental results are taken from a simulation and we present sample outcomes that show promising results.

1.1 Related work

Generally, our work belongs to the field of distributed algorithms for robot swarms, where groups of small, cheap robots are used to perform a variety of tasks. Some of these tasks are the formation of geometric patterns, gathering at and convergence to a single point, searching and partitioning the group of robots. A strategy for gathering which is closely related to ours is analyzed in [2]. The setting here also consists of robots which move synchronously, are oblivious and anonymous and have a limited viewing radius. Also, like in our strategies, the calculation of the next position of a robot is based on the center of the smallest enclosing circle around its neighbors. The main difference to ours is the goal of the strategy: there are no fixed stations and the robots gather in one point. Moreover, no robots are deleted and the avoidance of connection loss is different. In [2], the robots are restricted to a movement distance of at most 1 which must be calculated each round by each robot based on its neighborhood.

Another related scenario was presented in [3], where the explorer and the base station in the static setting correspond to our stations. In addition to only using two stations, a communication chain already exists such that every relay knows its predecessor and successor. Also in that scenario, removal of relays is not considered. The only strategy with removals is presented in [4]. It converges very fast, but also only holds for a chain of relays connecting two stations. In addition, the relays need a sense of direction, the chain has to be consistently directed from one station to the other.

1.2 The model

In the following, we will refer to both relays and stations together as nodes. Using a communication distance of 1, the unit disk graph of the stations and relays is a graph where vertices correspond to stations or relays and edges exist between two vertices if and only if the two nodes are within communication distance of each other. As stated in Section 1, we assume that the unit disk graph is connected at the beginning, and our goal is to transform the initial unit disk graph into one which still connects all stations and uses as few relays as possible. Note that an optimal selection would place the relays on the edges of a Minimum Steiner Tree connecting the stations, so that neighboring nodes have distance 1. It is assumed that the measurement of the position of neighbors is exact.

To simplify the analysis, we use a synchronous time model with discrete time steps. In each step, nodes are able to sense their environment within the communication distance of 1, to compute their new positions and to move toward this new position up to a distance of 1. This model is called "LCM-Model" (Look-Compute-Move Model) in [5]. We furthermore denote the position of robot r at the beginning of time step t by p_r^t and the set of neighbors of robot r at the beginning of time step t, i.e. all nodes within communication distance 1, by $N^t(r)$.

We assume that a relay i can be deleted from the system if it recognizes that it is not needed. The exact notions of "not needed" are part of the definition of our strategies. (A deleted relay may be assumed to return to some base camp, in order to be used for another task.)

For our analysis of the number of relays finally needed for the communication network, we need the technical assumption that the viewing radius is smaller than 1, i.e. we assume that the disks defining the unit disk graph are open sets. In this case, the initial unit disk graph has the property that connected nodes are in distance of at most $1 - \delta$, for some $\delta > 0$. We refer to δ as the *slackness* of the initial configuration. In case we allow closed disks, we say that there is no slackness.

2 The strategies

The following GO-TO-THE-CENTER strategy is executed sequentially, i.e. every time step one relay performs the strategy. The order in which the relays act is given from the beginning and does not change. The time from the beginning of a move of a relay r to the beginning of its next move is called a *run*. (We note here that our results also hold if we assume a parallel execution of the actions of relay, as long as no relays which are neighbors move in parallel. We have implemented a randomized such strategy.) If relay i performs the strategy in time step t, it first observes the exact positions of all its neighbors. It then computes its new position as *the center of the smallest enclosing circle around all positions of its neighbors within its viewing radius*. This center is equivalent to the point that minimizes the maximum distance to the nodes in $N^t(i)$. If there already is a node j at this position, relay i deletes itself.

The EXT-GO-TO-THE-CENTER strategy is an extension of the GO-TO-THE-CENTER strategy with an additional deletion rule. Here, before a relay i moves, it checks if the subgraph of the unit disk graph induced by $N^t(i)$ is connected. If this is true, i deletes itself.

The following lemma shows that the unit disk graph stays connected in both strategies.

Lemma 1. *For both strategies, the following holds: If the unit disk graph is connected before time step t, it is also connected at the end of time step t.*

Proof. Let relay i and j be neighbors at the beginning of time step t. At most one of the two relays can change its position in time step t. If none of them moves, they are obviously neighbors at the end of time step t. Let i therefore be the relay performing one of the two strategies. The point computed by i minimizes the maximum distance to its neighbors, therefore the maximum distance between i and its neighbors cannot increase:

$$d_{t+1}(i,j) \le \max_{k \in N^t(i)} \{d_t(i,k)\} \le 1$$

If i is not deleted, i and j therefore remain within communication distance and the unit disk graph remains connected. If i is deleted, there are two possibilities:

(1) There is another relay r at position p_i^{t+1} with $N^{t+1}(i) = N^{t+1}(r)$. i and r therefore have the same neighbors in the unit disk graph at the beginning of time step $t+1$ and all edges adjacent to i lie in a circle. If i is deleted, the unit disk graph remains connected.

(2) Relay i recognizes that its neighborhood is still connected without i. Since the deletion of i only affects its neighborhood, the whole graph remains connected. $\square$

3 Theoretical insights

This section deals with the worst-case analysis of our strategies. We first prove that, after finite time, the unit disk graph does not change any more, and that the convex hull of all nodes converges to the convex hull of the stations. Then we provide upper and lower bounds for the number of relays for scenarios with two or more stations for both strategies. In the following, we assume that the number of relays in the start configuration is n and that there exist m stations. The center of a smallest enclosing circle is defined by two or three points (f. ex. [6]) called basis, we also call the nodes at these positions the basis of a relay r.

Theorem 1. *The number of changes of the unit disk graph is bounded by $O(n(n+m))$.*

Proof. As soon as two relays i and j are neighbors, they stay neighbors until i or j is deleted. A change of the unit disk graph therefore is a result of the deletion of a relay or of two relays becoming neighbors. As no relays are added to the system, the maximum number of relays which can be deleted is n. Moreover, every relay can only find $n+m-2$ new neighbors, resulting in a maximum number of changes of the unit disk graph caused by new neighbors of $\frac{n}{2}(n+m-2) = \frac{1}{2}n^2 + \frac{1}{2}nm - n$. The maximum number of changes of the unit disk graph in general is therefore $\frac{1}{2}n^2 + \frac{1}{2}nm$. □

Theorem 2. *All relays will eventually be within ε-distance to CH^*, the convex hull of the stations, for every $\varepsilon > 0$.*

The proof follows a similar concept of a proof in [2], where it is shown that the robots gather in one point.

Proof.
Let $CH(t)$ denote the convex hull of all nodes after t steps. We split the proof in the following two parts. They obviously imply the theorem.

(a) For every $t \geq t_0$, $CH(t+1) \subset CH(t)$.
(b) There exists no convex polygon CH s.th. CH^* is a proper subset of CH and CH is a subset of every $CH(t)$.

Proof of (a):
Fix some time step $t+1$, in which some node i moves. This move transforms $CH(t)$ to $CH(t+1)$. As i's new position is contained in the convex hull of its neighbors, it is also contained in $CH(t)$, which implies (a).
Proof of (b):
Let CH be such a convex polygon. Furthermore let t_0 be a time instant such that for all $t \geq t_0$ all relays are within ε distance of CH for $\varepsilon > 0$ small enough and such that the unit disk graph does not change any more within finite time. Let CH' be another convex polygon consisting of edges parallel to those of CH in distance ε outwards from CH. CH must have a corner e which is not defined by a station, since CH^* is a proper subset of CH. We call the corresponding corner of CH' e'. Then there exists a line l intersecting CH in distance c of e with equal angles to both edges of CH that intersect in e, where $c > 0$ is sufficiently small (compare Figure 1). Let E be the isosceles triangle formed by l and the edges of CH' intersecting in e'. For every $t > t_0$, E must contain a set of relays R_E which can change in every time step, since otherwise the convex hull of all nodes would be a subset of CH (contradiction to the assumption). As the unit disk graph of the stations and relays is connected, at least one relay in R_E lies within communication distance to a relay r outside of E. W.l.o.g. let r be the relay with maximum distance to l which still has a neighbor in E. By choosing ε and c appropriately, it can be guaranteed that the distance d' from r to l is large enough such that in the run directly after r's move all relays in R_E within communication distance of r leave E. Let r_i be the first relay leaving E in this run. Then there must still be relays left in E after r_i's move and, since these relays are r_i's neighbors, the new position of r_i must be closer to l than the position of r. With ε and c small enough, the whole triangle E now lies within communication distance of r_i. Moreover, the distance of r_i to l is still large enough so that the new positions of all neighbors of r_i are outside of E. This means that in

the following run before r_i's next move all neighbors of r_i leave E or cannot enter it. Furthermore, every other relay entering E in this run would become neighbor of r_i contradicting the assumption. So, on r_i's next move, there are no nodes in E. This is a contradiction to the assumption. □

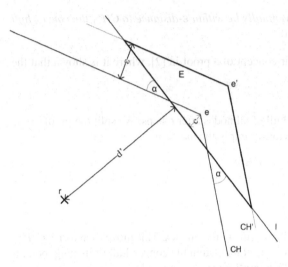

Figure 1 Proof of Theorem 2

In the following, we prove first upper and lower bounds on the number of relays eventually left over by our strategies. The first bound shows worst case limitations of our strategies.

Theorem 3. *Consider four stations forming a square of edge length d. There are configurations of $\Theta(d^2)$ relays for both strategies so that no relay will ever be removed. (Note that the minimum Steiner Tree has size $\Theta(d)$ in this case.)*

The second bound is restricted to the scenario of two stations and shows the asymptotic optimality in this scenario for initial configurations with positive slackness.

Theorem 4. *Let an initial configuration with two stations in distance d and some positive slackness δ be given. The number of relays used by EXT-GO-TO-THE-CENTER will eventually be at most 2d. (Note that this is optimal up to a factor of 2.)*

Proof (of Theorem 3). If the relays form a grid of distance 0.9 inside the square of stations, there exist edges in the unit disk graph between two adjacent points on the grid (see Figure 2). This is why there exists no relay with a connected neighborhood. Moreover, every relay is positioned in equal distance to all its neighbors in the center of the smallest enclosing circle. Therefore no relay can move or be deleted. The number of relays is then quadratic in d.

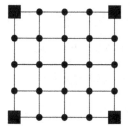

Figure 2 Illustration for the proof of Theorem 3

Proof (of Theorem 4). We first note, that the slackness of a configuration will not decrease anymore after the last of the (finitely many, compare Theorem 1) changes of the unit disk graph. Let the slackness after this last change be $\delta > 0$. Now consider a later time step so that, from now on, the relays will always stay within a rectangle around the line connecting the stations with width $\varepsilon > 0$ small enough so that $\varepsilon(1 - \delta) < 1$. Fix an arbitrary relay i. Because of the choice of ε, its neighbors in direction to the first station form a connected subgraph of the unit disk graph, and its neighbors in direction to the other station do so as well. Thus, as the unit disk graph does not change anymore, the closest neighbors of i in both directions have distance at least 1. This implies the bound $2d$ for the number of relays. □

4 Experimental results

In this section we will present some preliminary experimental simulation results. In our tests, the results were much better than the worst-case scenarios in Section 3. We have chosen three sample start configurations and present the results for both strategies.

In the first start configuration (Figure 3), the four stations form a square with side length 2.5 and part of the relays are located around the stations in form of a square with side length 5. The remaining of the 53 relays are situated arbitrarily inside or close to the square of relays.

The second start configuration (Figure 5) consists of five stations and 70 relays with the stations situated arbitrarily in the plane and a lot of relays in the convex hull of the stations. The remaining relays are mainly positioned at the left and at the upper right side of the convex hull of the stations and the longest distance between two stations is 6.

The third start configuration (Figure 7) consists of five stations and 400 relays, all of them positioned uniformly at random in a 10×10-Grid. Because of the big number of relays, the probability for a connected unit disk graph or for connected components which merge during the simulation is high.

These preliminary experiments suggest the following:

- The set of relays is thinned out considerably.

- The final formation of the relays does not reach the shape of a Minimum Steiner Steiner Tree, but comes close.
- Typically, if the formation contains a part that resembles a cycle within the convex hull of the stations, then this cycle cannot be broken by our local strategy.
- EXT-GO-TO-THE-CENTER converges substantially faster than GO-TO-THE-CENTER.

Figure 4 shows the configurations after applying GO-TO-THE-CENTER or EXT-GO-TO-THE-CENTER resp. to the first start configuration, Figure 6 depicts the results for the second and Figure 8 for the third start configuration. Additionally, the figures show the minimal Steiner tree as the globally optimal solution.

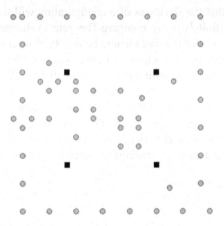

Figure 3 Start configuration for scenario 1

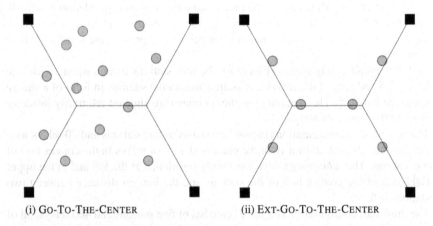

(i) GO-TO-THE-CENTER   (ii) EXT-GO-TO-THE-CENTER

Figure 4 End configurations for scenario 1

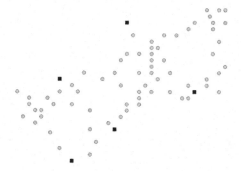

Figure 5 Start configuration scenario 2

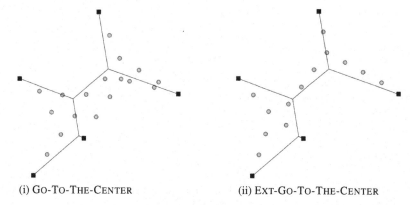

(i) Go-To-The-Center   (ii) Ext-Go-To-The-Center

Figure 6 End configurations for scenario 2

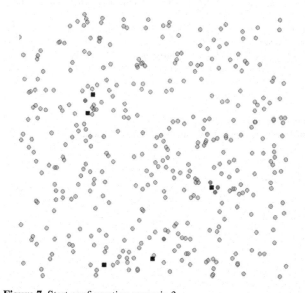

Figure 7 Start configuration scenario 3

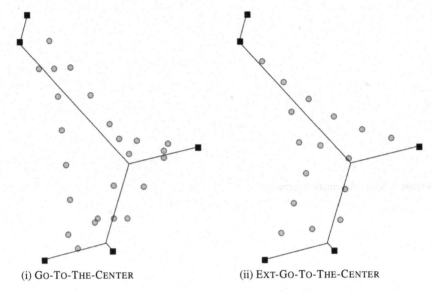

(i) GO-TO-THE-CENTER   (ii) EXT-GO-TO-THE-CENTER

Figure 8 End configurations for scenario 3

References

1. Y. U. Cao, A. S. Fukunaga, and A. B. Kahng, "Cooperative mobile robotics: Antecedents and directions," in *Autonomous robots*, vol. 4, pp. 1–23, 1997.
2. H. Ando, I. Suzuki, and M. Yamashita, "Formation and agreement problems for synchronous mobile robots with limited visibility," in *Proc. IEEE Int. Symp. Intelligent Control*, pp. 453–460, 1995.
3. M. Dynia, J. Kutylowski, P. Lorek, and F. Meyer auf der Heide, "Maintaining communication between an explorer and a base station," in *Proc. of the 1st IFIP Int. Conf. on Biologically Inspired Cooperative Computing (BICC)*, IFIP, pp. 137–146, Springer-Verlag Berlin, 2006.
4. J. Kutylowski and F. Meyer auf der Heide, "Optimal strategies for maintaining a chain of relays between an explorer and a base camp." To appear in Theoretical Computer Science, 2008.
5. R. Cohen and D. Peleg, "Robot convergence via center-of-gravity algorithms," in *Proc. of the 11th Int. Colloq. on Structural Information and Communication Complexity (SIROCCO)*, vol. 3104 of *Lecture Notes in Computer Science*, pp. 79–88, 2004.
6. S. Skyum, "A simple algorithm for computing the smallest enclosing circle," vol. 37, no. 3, pp. 121–125, 1991.

Measurement of Robot Similarity to Determine the Best Demonstrator for Imitation in a Group of Heterogeneous Robots

Raphael Golombek, Willi Richert, Bernd Kleinjohann, and Philipp Adelt

Abstract Imitation is not only a powerful means to drastically downsize the exploration space when learning behavior. It also helps to align the learning efforts of a robot group towards a common goal. However, one prerequisite in imitation, the decision of which robot to imitate, is often factored out in current research.

In our work we address this question by providing a means to measure the similarity between two robots. Based on this similarity a robot can choose which robot to imitate. The affinity of two robots with respect to imitation is most reasonably measured by calculating their behavioral difference, since the goal of imitation is learning new behavior. This is accomplished by each robot individually constructing an Affordance Network which is a Bayesian network upon its conditional affordance probabilities in the environment. An affordance represents the interaction possibilities an object provides to the robot. These Affordance Networks are then compared with a new metric.

1 Introduction

Imitation is not only a powerful means to drastically downsize the exploration space when learning new behavior [2, 12, 17]. It also helps to align the learn efforts of a robot group towards a common goal. It becomes especially important if a robot is a member of a group of robots who have to accomplish a common task or even several different tasks. The awareness of this has led to the definition of the "big five" questions in imitation, "namely who, when, what, and how to imitate, in addition to the question of what makes a successful imitation" [5]. As recent research has concentrated on the "what" and "how", the "who" has so far been factored out in current research – either by restricting the imitation process to a one-to-one demonstrator-imitator relationship where the roles of both are clear, or by providing the robots with fixed rules. However, the question of whom to actually imitate plays a role al-

Intelligent Mobile Systems, University of Paderborn / C-LAB, Germany, `richert@c-lab.de`

Please use the following format when citing this chapter:

Golombek, R., Richert, W., Kleinjohann, B. and Adelt, P., 2008, in IFIP International Federation for Information Processing, Volume 268; *Biologically-Inspired Collaborative Computing*; Mike Hinchey, Anastasia Pagnoni, Franz J. Rammig, Hartmut Schmeck; (Boston: Springer), pp. 105–114.

ready in early childhood, as shown e.g. by the Psychologist Burnstein [3]: He found out that children imitate more often peers that have similar sex, age or interests.

In our work we address this question by providing a means to measure the similarity between two robots. The affinity of two robots with respect to imitation is most reasonably measured by calculating their behavioral difference, since the goal of imitation is learning new behavior. This is accomplished by each robot individually constructing an Affordance Network which is a Bayesian network upon its conditional affordance probabilities in the environment. Encoded in such a network is the information which capabilities are dependent on which other ones. If, e.g. a robot knows it has the capability A and another robot is capable of A and B and in addition has identified the dependency $A \rightarrow B$, then it would be wise to imitate that robot. Two robots can thus individually learn about each other's Affordance Networks and calculate the difference. When comparing each other the robots can then determine from this difference the degree of behavioral similarity – the more behaviorally similar two entities are the more reasonable it would be for them to mutually imitate beneficial behavior. Depending on the metrics used to calculate the difference an observing robot can decide whether it should a) copy the knowledge of the other robot (if the demonstrator has the possibilities to share it), b) imitate the robot using indirect observation, or c) ignore it.

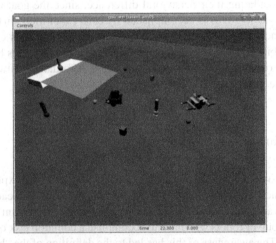

Fig. 1 Simulation environment with two morphologically different robots: Choosing the right robot to imitate helps to avoid useless imitation attempts

2 Related Works

The psychologist Gibson observed that our perception of the world is dependent on our interactions with it. For this he introduced the term *affordance* [9, 10] which is a property an object can have that describes the possible actions that can be performed with it. This depends on the one hand on the object itself, its objective properties like size, weight, surface friction, or shape, and on the other hand on the entity that tries to manipulate that object. But even if the system has all the capabilities to manipulate an object it does not help until it knows how to do that – a common learning problem in developmental robotics [13]: If a system is able to find out which actions make generally sense, it has filtered out the vast amount of useless actions.

Robotics researchers have embraced that concept of affordances as it helps the research to look through the eyes of a robot [7, 15, 16]. Affordances are even used together with Bayesian Networks (BNs) in the field of imitation: Lopes et al. [14], e.g., use BNs to learn object affordances. However, as they are interested in learning the individual affordances they assume the role of the demonstrator to be known. Thus they use BNs to model the affordance. In contrast, we use them to model affordance dependencies to infer behavioral differences. Cesa-Bianchi et al. [4] present an algorithm that chooses the best expert from a set of predefined experts. In their game-theoretic approach they require a static set of always accessible experts – a condition hardly met in realistic robotic domains. As their algorithm is only relying on the experts' performances it does not help in heterogeneous real-world robotics applications, where robots have different morphologies and capabilities. Balch developed a means to measure the overall diversity of a group of robots [1]. In his approach the robots are assumed to be morphologically similar and that their learning algorithms already have converged to a stable behavior.

3 Algorithm

To compare robots based on their behavioural affinity and thus actively control the imitation process one needs data which is related to the robot's behavioural possibilities. Upon the raw data we build a meaningful representation which we then use to compare robots by an adequate metric.

3.1 Gathering Raw Data

First we need data about behavioural possibilities of a robot. This information relies on the robot's hardware and software. Thus the first idea could be to compare these components. However, this approach is infeasible because the information can be exchanged by communication only and a common communication interface cannot be demanded for arbitrary robots. Furthermore, the robot's hardware differs even if it fulfills the same functionality, i.e. one robot could be differentially driven while another one uses omniwheel for locomotion.

As already pointed out, affordances are subjective and tightly coupled to the behavioural capabilities of a robot. Each robot can gather its own affordance data during environmental exploration and can get the affordance data of potential demonstrators by observation. In this paper we assume the set of affordances to be predefined and constant. However, this is no restriction to our algorithm as learning affordances at run-time (e.g. via [14]) does not pose a problem to this approach. Let $S = \{s_1, ..., s_n\}$ be the set of all recognizable affordances and let $L = \{l_1, ..., l_n\}$ denote all objects in the robot's environment. We can then define $O = S \times L \times B$ where $B = \{True, False\}$ as the set of all possible results of an affordance test. Furthermore, we define $O_{ri} \subseteq O$ as the set of results gathered by robot R_i. To simplify the disscussion we define the following functions: $bool(o_{ir}) = b_{ir}$, $affordance(o_{ir}) = s_{ir}$, and $object(o_{ir}) = l_{ir}$, whereas $o_{ir} \in O_{ir}$.

3.2 Affordance Network

The gathered raw data is unstructured, noisy and incomplete. Furthermore, the set of data samples grows rapidly during the robot's environmental exploration. To structure the data and cope with uncertainty each affordance is interpreted as a random variable X_i and the gathered data for this affordance as a sample set. We then can define the finite set $X = \{X_1, ..., X_n\}$ of random affordance variables where each variable may take on a value x_i from the domain $\{True, False\}$. Upon O_{ri} we define $O'_{ri} = \{(bool(o_{r1}), ..., bool(o_{rm})) \mid \forall o_{ri}, o_{rj} : i \neq j$ and $object(o_{ri}) = object(o_{rj})\}$ and interpret these tuples as samples of the joint distribution of the random variables in X. To get a compact representation of the joint distribution of the variables in X we train a Bayesian Network with the set O'_{ri} for each robot individually. As the data in O'_{ri} is directly coupled to the tested affordances a Bayesian Network trained with this set will also encode behavioral information.

Definition 1. Let P be a joint probability distribution of the random variables $X = \{X_1, ..., X_n\}$ in some set V, and $G = (V, E)$ be a DAG. We call (G, P) a Bayesian Network if (G, P) satisfies the Markov condition. By applying the chain rule of probabilities and properties for conditional independencies, any joint distribution P that satisfies the Markov condition can be decomposed into the product form:

$$P(X_1, ..., X_n) = \prod_{i=1}^{n} P(X_i \mid parents(X_i)).$$

The directed edges of the DAG describe causal relations between the random variables in X. Each node has an attribute which describes the conditional probabilistic distribution of its random variable and the random variables of its parents.

If there is enough expert-knowledge to define the structure of a Bayesian Network, only the parameters i.e. the conditional probabilities have to be learned from data. In our case as we do not use any further domain knowledge we have to learn both, the structure and the parameters from data. Therefore we apply the Structural

EM Algorithm [8] to the affordance data. This is an iterative algorithm based on a standard Expectation Maximization algorithm to optimize parameters, and a structure search to find the current best structure model.

A problem commonly found in structural-learning Bayesian networks is that real causality cannot be derived from raw data [11]. However, this is not a problem here as we do not make inference on the trained networks rather we use them to measure the distance between robots by means of behavioral affinity.

3.3 Metric

After defining a meaningful and well structured representation of a robot we now need to define a metric to measure robot affinity. As BNS are directed acyclic graphs we can apply the Graph Edit Distance Metric (GED) [6] to measure structural distance between two graphs g_1 and g_2. To describe the GED metric we need the definition of a label representation as defined in [6]:

Definition 2. Let L_E and L_V denote sets of edge and node labels, respectively. A graph $g = (V, E, \alpha, \beta)$ is a 4-Tuple where V is the finite set of vertices, $E \subseteq V \times V$ is the set of edges, $\alpha : V \rightarrow L_V$ is a function assigning labels to the nodes and $\beta : E \rightarrow L_E$. The label representation of g, $p(g)$, is given by $p(g) = (L, C, \lambda)$:

- $L = \{\alpha(x) | x \in V\}$,
- $C = \{(\alpha(x), \alpha(y)) | (x, y) \in E\}$, and
- $\lambda = C \rightarrow L_E$ with $\lambda(\alpha(x), \alpha(y)) = \beta(x, y)$ for all $(x, y) \in E$.

Using the label representation we can then define the Graph Edit Distance metric [6]:

Definition 3. Let g_1, g_2 be two graphs with label representations $p(g_1)$ and $p(g_2)$. Furthermore, let $C_0 = \{(i, j) | (i, j) \in C_1 \cap C_2 \text{ and } \lambda_1(i, j) = \lambda_2(i, j)\}$ and $C_0' = \{(i, j) | (i, j) \in C_1 \cap C_2 \text{ and } \lambda_1(i, j) \neq \lambda_2(i, j)\}$. Then the graph edit distance $d(g_1, g_2)$ of the two graphs is

$$d_{ged}(g_1, g_2) = |L_1| + |L_2| - 2|L_1 \cap L_2| + |C_1| + |C_2| - 2|C_0| + |C_0'| \qquad (1)$$

As we are only interested if an edge between two nodes exists, we define $\beta(x, y) = 1$ $\forall (x, y) \in E$, and thus omit $|C_0'|$ from our distance metric for affordance networks. As the nodes of the Affordance Networks also need a special treatment due to the inherent conditional probability differences, this leads to the edge comparing part of the final affordance network metric d_{ane}:

$$d_{ane}(g_1, g_2) = |C_1| + |C_2| - 2|C_0| \qquad (2)$$

The nodes of the Affordance Network contain conditional probabilities which may differ so that we have to measure distances between these probabilities. For example, consider the node with the label "Reachable" in the networks g_1 and g_2

(Fig. 2). In g_1 it has no parent, thus it is is said to be unconditioned. The node with the same label in g_2 is conditioned by the outcome of node with the label "Pushable". We have to compare $P(Reachable) = 0.2$ from the node in g_1 with $P(Reachable|Pushable = 1) = 0.0$ and $P(Reachable|Pushable = 0) = 0.8$ from the node in g_2. To be able to measure the distance inside the nodes we use the definition

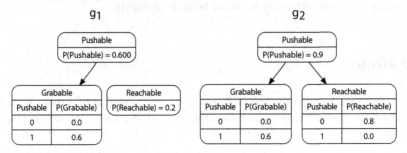

Fig. 2 Two Affordance Networks with different distributions

of independence and conditional independence.

Definition 4. Two events E and F are independent if one of the following holds:

- $P(E|F) = P(E) \wedge P(E) \neq 0, P(F) \neq 0$
- $P(E) = 0 \vee P(F) = 0$

Definition 5. Two events E and F are conditionally independent given an event G if $P(G) \neq 0$ and one of the following holds:

- $P(E|F \cap G) = P(E|G) \wedge P(E|G) \neq 0, P(F|G) \neq 0$
- $P(E|G) = 0 \vee P(F|G) = 0$

Using these definitions we can expand the probability of the node in g_1: The probability for the "Reachable" affordance is therefore transformed from $P(Reach.) = 0.2$ to $P(Reach.|Push.) = 0.2$ and $P(Reach.|\neg Push.) = 0.2$ if the events "Reachable" and "Pushable" are independent. Since there is no edge in g_1 between the nodes with labels "Reachable" and "Pushable" the Markov condition guarantees their independence.

After extending the probability labels we can interpret the probabilities of node v as a point $point(v)$ in n-dimensional space, where n is the number of entries in the nodes probability table. Then we can calculate the distance δ (e.g. Euclidean) between two equally labeled nodes in the different Affordance Networks:

$$d_{ann}(g_1, g_2) = \sum_{\substack{v_1 \in V_1, v_2 \in V_2 \\ \alpha(v_1) = \alpha(v_2)}} \delta(point(v_1), point(v_2)) \tag{3}$$

The final distance function for two Affordance networks d_{an} is then the weighted summation of $d_{ane}(g_1, g_2)$ and $d_{ann}(g_1, g_2)$, where the weights are domain dependent and can be used to control the influence of the structure and the probability distribution:

$$d_{an}(g_1, g_2) = c_e \left(|C_1| + |C_2| - 2|C_0| \right) + \tag{4}$$

$$c_n \left(\sum_{\substack{v_1 \in V_1, v_2 \in V_2 \\ \alpha(v_1) = \alpha(v_2)}} \delta(point(v_1), point(v_2)) \right) \tag{5}$$

4 Experimental Results

The presented approach will be demonstrated with two scenarios: In the first one it is shown in detail how the approach leads to the determination of behavioral difference. The second scenario demonstrates how its usage leads to a significant improvement of the imitation process.

4.1 Scenario 1

In this artificial example there are three robots of which one is the imitator (R_i) that has to choose between two demonstrators (R_{d_1}) and (R_{d_2}) to imitate. The properties regarding the gripper and the drive motor are shown in Tab. 1. Fig. 1 shows the two demonstrators to which the environmental objects have different affordances because of their different morphologies: The yellow robot (R_{d_2}) with the barbed gripper, e.g., is able to pull objects and can not lift them, whereas the blue one (R_{d_2}) has a strong gripper but a weak drive, so that it is able to lift some objects, but can not pull them.

Over the course of its lifetime the imitator has recorded the affordances of the two other robots in the scenario and has built an Affordance Network as described in Sec. 3.2 that is depicted in Fig. 3 (page 8). Let us now take a look on how the presented behavioral metric works on those networks to measure the behavioral similarity.

Table 1 Qualitative description of three robots. The imitator is more similar to demonstrator 1 in terms of its gripper and motor capabilities. When imitating another robot in order to learn new behavior it should imitate that robot instead of demonstrator 2.

robot	capabilities			
		gripper		motor
	length	strength	style	strength
demonstrator 1	long	weak	barbed	strong
demonstrator 2	short	strong	normal	weak
imitator	normal	weak	barbed	normal

From the qualitative description it is intuitively clear in this simple example that the imitator has more resemblance to demonstrator 1 and should imitate that robot instead of demonstrator 2 (R_{d_2}). Applying the distance metric d_{an} to the data collected with the three robots we get the results for various weights of the edge and the node distance part as shown in Tab. 2. As can be seen the behavioral distance between the imitator and the first demonstrator is smaller than the distance to the second one.

Table 2 Behavioral similarity calculated using the distance metric d_{an}.

c_e	c_n	$d_{an}(R_i, R_{d_1})$	$d_{an}(R_i, R_{d_2})$
0.1	0.9	2.307	3.614
0.25	0.75	2.423	3.845
0.5	0.5	2.61	4.23
0.75	0.75	2.808	4.615
0.9	0.1	2.923	4.846

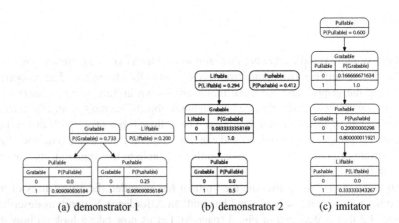

(a) demonstrator 1   (b) demonstrator 2   (c) imitator

Fig. 3 The final Affordance Networks from the viewpoint of the imitator. These are used by the imitator to calculate the difference in order to determine that robot that has behaviorally the most resemblance to the imitator

4.2 Scenario 2

This scenario was carried out in the PlayerStage/Gazebo simulation environment (Fig. 1). A robot similar to the well-known Pioneer2DX had to choose between three morphologically different demonstrators whom to imitate. All robots differed in the size, strength and shape of their gripper and the strength of their drive unit.

The experiment was carried out as follows: The imitator started without any knowledge and no Affordance Networks. Then it observed a random demonstrator carrying out an action like pushing an object (Fig. 1) and recorded whether it was successful. Afterwards it carried out all known tasks with all known objects in the environment by itself, recorded the success and updated the Affordance Networks for them both. The number of failed behaviors dependent on the number of imitations it has carried out is shown in Fig. 4. As can be seen the imitation process

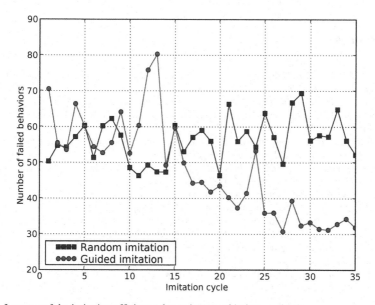

Fig. 4 Increase of the imitation efficiency due to improved imitator selection

gets more efficient with more observations and thus more exact Affordance Networks compared to randomly choosing a demonstrator. The Affordance Networks together with their metric significantly help to improve the overall imitation process.

5 Conclusion

We introduced Affordance Networks, which are Bayesian Networks based on the affordance dependencies of robots and developed a distance metric which calculates the behavioral diversity between two robots based on those Affordance Networks. Once the affordances are observed a robot willing to imitate is able to determine that robot in a heterogeneous robot group that is the most similar to the imitator. This leads to a higher success probability in the imitation process as imitation of robots that are e.g. morphologically different to the imitating robot is avoided.

Although this example is carried out only in realistic simulation environment it should be clear that it would result in a similar outcome in a real world if the data can be collected in sufficient quantity, because the presented approach is able to cope with missing and noisy data. To our knowledge this is the first solution to answer the question whom to imitate in a group of robots. It is not restricted to the robotics domain, but can be applied to all domains where behavior can be observed and imitated.

References

1. Balch, T. *Behavioral Diversity in Learning Robot Teams*. PhD thesis, Georgia Institute of Technology, Dec. 1998.
2. A. Billard and M. J. Mataric. Learning human arm movements by imitation: : Evaluation of a biologically inspired connectionist architecture. *Robotics and Autonomous Systems*, 37(2-3):145–160, 2001.
3. E. Burnstein, E. Stotland, and A. Zander. Similarity to a model and self-evaluation. *Journal of Abnormal and Social Psychology*, 62:257–264, 1961.
4. N. Cesa-Bianchi, Y. Freund, D. Haussler, D. P. Helmbold, R. E. Schapire, and M. K. Warmuth. How to use expert advice. *J. ACM*, 44(3):427–485, 1997.
5. K. Dautenhahn and C. Nehaniv. An agent-based per- spective on imitation, 2002.
6. P. J. D. et. al. On graphs with unique node labels. In *Graph Based Representations in Pattern Recognition*, volume 2726, pages 409–437, Heidelberg, DE, 2003. Springer Berlin.
7. P. Fitzpatrick, G. Metta, L. Natale, S. Rao, and G. Sandini. Learning about objects through action-initial steps towards artificial cognition. *Robotics and Automation, 2003. Proceedings. ICRA'03. IEEE International Conference on*, 3, 2003.
8. N. Friedman. Learning belief networks in the presence of missing values and hidden variables. In *Proc. 14th International Conference on Machine Learning*, pages 125–133. Morgan Kaufmann, 1997.
9. J. Gibson. The theory of affordances. In R.Shaw and J.Brandsford, editors, *Perceiving, Acting, and Knowing: Toward and Ecological Psychology*, pages 62–82. Erlbaum, Hillsdale, NJ, 1977.
10. J. J. Gibson. *The Senses Considered as Perceptual Systems*. Houghton-Mifflin Company, Boston, 1966.
11. D. Heckerman, D. Geiger, and D. M. Chickering. Learning bayesian networks: The combination of knowledge and statistical data. *Mach. Learn.*, 20(3):197–243, 1995.
12. A. Ijspeert, J. Nakanishi, and S. Schaal. Movement imitation with nonlinear dynamical systems in humanoid robots, 2002.
13. M. Kopicki, A. Sloman, J. Wyatt, and R. Dearden. Learning object affordances by imitation. Technical report, The University of Birmingham, 2005.
14. M. Lopes, F. Melo, and L. Montesano. Affordance-based imitation learning in robots. In *2007 IEEE/RSJ International Conference on Intelligent Robots and Systems*, 2007.
15. E. Oztop, N. Bradley, and M. Arbib. Infant grasp learning: a computational model. *Experimental Brain Research*, 158(4):480–503, 2004.
16. A. Slocum, D. Downey, and R. Beer. Further Experiments in the Evolution of Minimally Cognitive Behavior: From Perceiving Affordances to Selective Attention. *From Animals to Animats 6: Proceedings of the Sixth International Conference on Simulation of Adaptive Behavior*, 2000.
17. A. Ude, T. Shibata, and C. G. Atkeson. Real-time visual system for interaction with a humanoid robot. *Robotics and Autonomous Systems*, 37(2-3):115–125, 2001.

Distributed Fault-Tolerant Robot Control Architecture Based on Organic Computing Principles

Adam El Sayed Auf, Marek Litza, Erik Maehle

University of Lübeck, Institute of Computer Engineering, Ratzeburger Alle 160, 23538 Lübeck, Germany
{elsayedauf, litza, maehle}@iti.uni-luebeck.de

Abstract Walking animals like insects show a great repertoire of reactions and behaviours in interaction with their environment. Moreover, they are very adaptive to changes in their environment and to changes of their own body like injuries. Even after the loss of sensors like antennas or actuators like legs, insects show an amazing fault tolerance without any hint of great computational power or complex internal fault models. Our most complex robots in contrast lack the insect abilities although computational power is getting better and better. Understanding biological concepts and learning from nature could improve our approaches and help us to make our systems more "life-like" and therefore more fault tolerant. This article introduces a control architectural approach based on organic computing principles using concepts of decentralization and self-organization, which is demonstrated and tested on a six-legged robotic platform. Beside explaining the organic robot control architecture, this study presents a leg coordination architecture extension to improve the robustness and dependability towards structural body modifications like leg amputations and compares experimental results with previous studies.

Introduction

Legged locomotion encounters challenges like controlling the leg's different phases with unequal responsibilities, coordinating legs to achieve an effective walking pattern as well as adapting the achieved gait to the environmental circumstances and a given task. In contrast to a wheeled rotary motion, a leg passes through a stance phase, carrying part of the systems weight and stemming it to a given direction, and a swing phase, lifting up the leg and moving it in the opposite direction to put it down again. The precondition for a stance phase is the leg's ground contact and must be obtained at the end of each swing phase. All legs of a

Please use the following format when citing this chapter:

Auf, A.E.S., Litza, M. and Maehle, E., 2008, in IFIP International Federation for Information Processing, Volume 268; *Biologically-Inspired Collaborative Computing;* Mike Hinchey, Anastasia Pagnoni, Franz J. Rammig, Hartmut Schmeck; (Boston: Springer), pp. 115–124.

walking system accomplishing alternating stance and swing phases must be coordinated to a stable walking pattern. In the animal kingdom, these tasks are solved by each single walking creature. Even insects that are alleged to have low "computational power" cope with these tasks. Animals use smooth transitions of their gaits instead of hard switches. Horses do not switch between standing on a feedlot, trotting and galloping but use a smooth changeover between different gaits depending on the animal's velocity, equally insects do. That leads to the assumption that controlling walking patterns does not base on a set of predefined gaits, but the use of general rules providing the ability of adaptation. Taking into account the mentioned challenges are passed by each walking animal, the question arose if a central or decentral control is underlying legged locomotion. The present work introduces an organic computing approach facing the challenges with decentralised and self-organizing mechanisms by the use of a six-legged walker.

Figure 1: The Robot Platform OSCAR.

The Robot Platform

OSCAR (Organic Self Configuring and Adapting Robot) is a six-legged walking machine with 18 degrees of freedom (DOF). Its servo motors as well as its skeleton parts are commercially available. Set together, the parts result in a symmetric robot with a round body and a span length of 74.5 cm over completely stretched out legs (Fig. 1). Each leg has three degrees of freedom and is equipped with a binary contact sensor. Its head consists of a wireless camera, an ultra sonic sensor and a heat sensor, altogether movable with one joint. While the binary contact sensor signals if the leg is touching the ground or not, the camera sends a color picture to a personal computer, the ultrasonic sensor measures the distance between OSCAR and an obstacle and the heat sensors detects heat sources like humans. The onboard control hardware, also commercially available, consists of a SD21 Servo Driver Module and a JControl/SmartDisplay, programmable in Java. All parts are connected by an I2C-Bus.

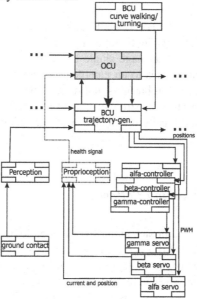

leg i

Figure 2: Example for the Organic Robot Control Architecture ORCA for one leg.

The Software Architecture

The Organic Robot Control Architecture - ORCA – was designed to be modular and hierarchically organized in order to be easily manageable and to fit to the control architecture of a robot which is also assumed to be modular, hierarchically organized and behaviour-based [1,2]. An ORCA-based system is built using Basic Control Units (BCUs) to achieve the desired functionality. Each BCU encapsulates a specific functionality. It can for example implement generic signal filtering modules or PID-controllers, but may also encapsulate sensor or actor hardware of the robot at lower system levels. BCUs interact by interchanging data (signals). The connections between BCUs can also be used to trigger activity in the receiving BCU when new data are sent. Organic Control Units (OCUs) supplement these BCUs. An OCU uses the same unified interface that BCUs use to interchange data and trigger activities. In contrast to BCUs, OCUs do not realize a predefined custom function for the robot, but they monitor the signals generated by one or more BCUs. When these monitored signals show a substantial change, the OCU can react by changing parameters of the BCUs. By defining "normal" or "good" ranges for some of the signals a BCU generates, an OCU can decide when to start changing parameters to bring the system back to a normal or "healthy" state (reasoner). Further, the BCUs shall be able to learn in order to improve their reactions in similar situations in the future by making use of a memory for a short-time history. So the BCUs manage single basic functions in close collaboration with other BCUs, which leads to an emergent collective functionality, while the OCUs ensure the system's robustness and dependability by monitoring single or several BCUs and reacting to stronger changes or failures. As well as BCUs, OCUs are modular and locally distributed with the ability to communicate with each other.

Self-Organizing Walking Patterns

Based on the ORCA approach and inspired by biological experiments and controllers like the Walknet [4] a decentralized controller is used to achieve organic walking patterns. Each of the six legs consists of three joints and has its own controller implemented as a separate BCU for trajectory generation (Fig.2). These BCUs generate position commands for the BCUs of the three leg joints (alpha, beta, and gamma-controller) which then send PWM (Pulse Width Modulation) signals to the respective servos.

Controlling one leg means managing an alternation composed of swing and stance phase. In the swing phase the leg is lifted up from the ground, moved to the front, and put down on the ground again. The ground contact signal triggers the stance phase when the leg moves backwards in respect to the body. In this phase it carries a part of the body's weight and pushes it to the front. To obtain an alternation between swing and stance phases, the trajectory controller uses the information of the ground contact signal and two extreme positions: the posterior extreme position (PEP) and the anterior extreme position (AEP). If a leg has reached the PEP in its stance phase, it switches into swing phase. So the leg will lift up and move along a fixed trajectory towards the AEP. After reaching the AEP and receiving the ground contact signal, the leg switches again into the stance phase and moves back towards the PEP. This principle leads to a swing-stance alternation. The frequency of this alternation can be changed by increasing or decreasing the duration of the stance phase only.

Each of the six trajectory BCUs uses the same flexible coordination rule to harmonize its movements with the neighbouring leg controllers' swing phases. The ground contact signals are processed by a perception BCU and then sent to the trajectory BCU which passes it to the trajectory BCUs of the two neighbouring legs. The leg controllers inhibit switching from the stance into the swing phase, when its two neighbouring BCUs are not sending a ground contact signal. This rule ensures that a maximum number of three legs are swinging at the same time. Thus, the robot's legs are always in a stable configuration. Combination of local alternating leg movements and the mentioned coordination rule lead to a global organic walking behaviour. Different walking patterns arise in a self-organizing way by increasing or decreasing the stance phase's duration without using a pre-programmed gait. In contrast to the stance phase the swing phase, as was observed in the stick insect Carausius morosus, has a fixed duration and can not be varied. So, the gait is adapting to the robot's walking velocity in a self-organizing way. In a very slow gait, achieved by usage of a stance phase about five times longer than the swing phase, a pentapod gait can be observed. So, five feet are on the ground while just one leg is swinging. Shortening the stance phase until it is as long as the swing phase, a tetrapod and a tripod result smoothly [5]. These gaits are observable in nature as well and biological experiments have shown that insects move in a very similar way and do not make use of a hard switching between their gaits.

Turning and Curve Walking

One advantage of OSRAR's symmetric body is its ability to rotate around its centre without taking more space than in a standing posture. In contrast to walking the legs have not to be separated in left and right legs, moving in opposite direction in respect to the body centre, during rotating all legs move into the same direction. A rotating of the robot is achieved by one BCU per leg setting the AEP and the PEP of each leg on the same value. So changing two parameters while maintaining the coordination rule leads to a completely different behaviour.

Another challenging behaviour is curve walking. One simple approach to achieve curve walking is to shorten the stance trajectories of the three legs at the inner side of the curve and enlarging the stance trajectories of the three legs at the outer side. Diverging the PEP and AEP of the outer legs and converging the PEP and AEP of the inner legs lead to a curve walking. Converging the PEP and AEP until they change sides and diverge again, leads in the end to the above described turning behaviour. Thus, one BCU per leg is sufficient to produce a smooth transition from straight walking over curve walking to turning by changing the AEP and PEP values of the legs.

Another approach of curve walking, which can also be observed in experiments with the stick insect Carausius morosus [3], is to change the trajectory of each leg independently. Further biological experiments and computer simulations showed that decentralized control architecture combined with local rules lead to curve walking based on individual stance trajectories for each leg without explicit calculations [6]. In the ORCA project a mixture of both approaches is used. For walking in a curve, the front leg's PEP and AEP are shifted in the above described way. This shift of the extreme positions is detected by the next posterior legs and leads also in a shift of their extreme positions. So a change in the front leg extreme positions leads to a shift in the middle leg extreme positions and that leads to a shift in the hind leg extreme positions. Thus, a change in the front legs is spread by local rules through the whole system. The intensity of the middle and hind leg shifts depend on the intensity of their anterior leg shifts. In the following experiments front leg extreme position shifts are caused by a BCU that guides the robot to walk towards a detected stimulus. In the presented experiments this stimulus is a heat source, which can be detected by the heat sensor at the robot's head.

The System's Fault Tolerance

To show the system's robustness and dependability a strong structural body modification was realised by a leg deactivation, which was achieved by clicking the affected leg upwards and holding it above the robot's body. So a leg amputation is simulated, whereas in contrast to a real amputation the leg's weight is still being carried by the walking machine.

A deactivated leg is recognised by an OCU because of the missing feedbacks like ground contact or servo feedback. An OCU detecting a non-functional leg will modify the responsible trajectory BCU in a way that it is channelling its neighbouring legs the ground contact signals through. So the next leg waiting for the ground contact signal of the defect leg will receive the ground contact signal of the next functional leg. Thereby, the self-organized coordination of the remaining legs is continued and the walking pattern adapts to the system's new structure [5].

Experiments and Results

In previous experiments with the above mentioned software architecture the robot's walking abilities were shown. Beside the self-organising walking pattern and the robot's capability to handle leg damages during straight forward walking with an emergent gait [6], the walking machine's curve walking was tested in [7].

In [7] was shown that during a leg amputation the robot is still able to navigate towards a given goal position without any modifications in the robot's control software. But, without any changes of the machine's walking behaviour it is hindered in its stability during curve walking by the missing leg. The robot needs up to twice as long for reaching its goal position as in a "healthy" state with all six legs [7].

Coordination Extension

To remain true to the decentralised control architecture on the one hand and improve the robots walking behaviour in a certain situation on the other hand, the leg observing OCU was modified. The OCU of an amputated leg gives additional information to its neighbouring legs' OCUs. A neighbour leg changes its swing start conditions to make the emerging gap caused by a leg amputation as small as possible. The leg is waiting for its neighbours to come as close as possible in their stance phase before starting its own swing phase. Its neighbours' positions are depending on their neighbouring legs' swing directions. To give an example (FIG. 3a), the robot's legs are numbered clockwise when regarded from above. The neighbour leg ready to swing is called leg i and its functional neighbours are called i-1 and i+1. In case that leg i-1 is swinging clockwise (cw) and leg i+1 is swinging anti-clockwise (acw) leg i will start its swing phase when leg i-1 and leg i+1 are near their AEPs. Another example with the opposite case is given in Fig. 3b. When the front left leg (FL) i+1 swings clockwise and the hind right leg (HR) i-1 swings anti-clockwise the hind left leg (HL) i has to wait until its neighbours are close to their PEPs. All four possible situations of the explained example are shown in Tab. 1.

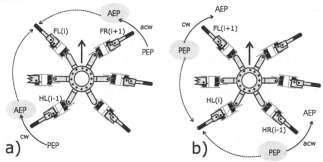

Figure 3: Example for the coordination extension. a) Front left leg (FL) as neighbour of a damaged leg is allowed to swing when its functional neighbours, front right leg (FR) and hind left leg (HL), are near their AEPs. b) Hind left leg (HL) as neighbour of a damaged leg is allowed to swing when its neighbours, front left leg (FL) and hind left leg (HL), are near their PEPs.

Table 1: Position areas in stance phase of the functional neighbours of leg i which are the preconditions for leg i to start its swing phase depending on the neighbouring leg's swing direction (anticlockwise acw or clockwise cw).

	i-1		i+1	
swing direction	acw	cw	acw	cw
area around	PEP	AEP	AEP	PEP

Experimental Setup

For being able to compare the experiments of this work to previous ones, the experimental setup is the same as in [7].
The setup is a 230x160 cm flat indoor-area in which the robot can freely walk. The Robot is equipped with two markers, one at its centre on top of his head and one at its front between the two front legs. The markers horizontal distance is 10 cm and the vertical 15 cm. A third marker is used for the heat source, also placed in the scene. The distance between the robot's centre marker and the position a foot touches the ground is varying depending on the robot's leg posture from 18 to 26 cm. The whole scenario is captured by a standard camera 310 cm above the ground. The results in [7] have shown the leg amputation's influence on the robot's curve walking behaviour. In two situations the robot is significantly hindered needs nearly twice as long for reaching its goal than in all other runs. These situations were a left curve walk with a right middle leg amputation and a right curve walk with a left middle leg amputation. In both cases the amputation causes a turning towards the hindered body side. The following experiments concentrate on these two critical situations. Two robot heat source configurations have been tested: one left middle leg amputation in a right curve walk and one right middle leg amputation in a left curve walk.

By rotating its head the robot scans its environment for heat sources. One of the results is shown in Fig.4. The coordination system in Figure 4 shows on the x and y-axes the robot's position in the indoor-area in cm. The robot's centre is marked as a circle with an arrow giving the robot's orientation, while the target heat source is represented as a diamond. Over time recorded positions of the robot result in the six-legged machine's walking path and are marked with additional timestamps. Time t=0 as well as the time passed until the robot reached its target and the time when the robot detects the heat source are given in Fig. 4. About the x-Position 120 the robot's right middle leg is amputated and at time 10 s the heat source is detected. Its orientation shifts towards the heat target while at about x-position 120 the robot's right middle leg was amputated to test the controller extension. Although the robot is still hindered by the missing leg that has to be carried with him, it reaches the heat source in a shorter time than in previous experiments. The time the hexapod reaches its target (63 s) is not comparable par to par because of the different distances that the robot walked in Fig.4 and Fig.5. Although the distance walked by the robot is longer in the previous runs (Fig.5), in three of five cases the walking machine reaches the heat source in a significant shorter time than in previous runs without the coordination extension. In the other two cases the robots does not find the heat source and misses its target.

Fig. 5 shows the result with the same setup and a robot using the not extended control architecture. In comparison Fig. 5 shows a stronger influenced walking path than it can be seen in Fig. 4. Using the control architecture without any additional information like it was done in previous works the robot reaches also its goal, but the above mentioned decentral extension to the control system, based on local rules between the neighbouring legs can improve the robots abilities as well as its robustness.

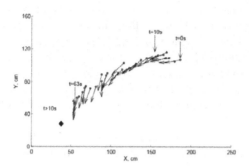

Figure 4: Robot's walking path shown as positions(XY). Robot's centre shown as circle, orientation shown as arrow, heat source as diamond. At time 10 s the robot detects the heat source and needs 63 s to reach the target. A right middle leg amputation [ra] was triggered at X-Position 120.

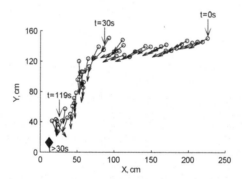

Figure 5: Result from a previous experiment when the robot uses the original control architecture without the extension. Robot's walking path shown as positions(XY). Robot's centre shown as circle, orientation shown as arrow, heat source as diamond. At time 30 s the robot detects the heat source and needs 119 s to reach the target. A right middle leg amputation [ra] was triggered at X-Position 120.

Summary

The introduced approach for controlling the six-legged walking machine OSCAR shows a complex curve walking behaviour with six legs handled by a decentralized control architecture based on local rules by changing a few parameters. Moreover, it is explained how continuous changing of parameters can induce a smooth transition between different walking behaviours and how a decentralized control architecture can be extended by additional local rules to improve the systems abilities. The presented experiment shows the improvement to the system's fault tolerance and robustness towards strong damages like leg amputations by a decentral extension of the control architecture and how the system is able to handle even strong damages while maintaining complex behaviours by adapting in a dependable way.

Beside the handling of strong damages it is open how to give the system the ability to detect smaller defects like loose screws or broken gears in its servo motors and if it is possible to compensate them.

Acknowledgments

This work was funded in part by the German Research Foundation (DFG) within priority programme 1183 under grant reference MA 1412/7-1.

References

1) Brockmann W, Großpietsch K.-E, Maehle E, Mösch F: ORCA - Eine Organic Computing-Architektur für Fehlertoleranz in autonomen mobilen Robotern. Mitteilungen der GI/ITG-Fachgruppe Fehlertolerierende Rechensysteme, Nr. 33, 3-17, St. Augustin 2006
2) Brockmann, W, Maehle E, Mösch F: Organic Fault-Tolerant Control Architecture for Robotic Applications. 4th IARP/IEEE-RAS/EURON Workshop on Dependable Robots in Human Environments, Nagoya University/Japan 2005
3) Dürr V, Ebeling W: The behavioural transition from straight to curve walking: kinetics of leg movement parameters and the initiation of turning. The Journal of Experimental Biology 208, 2237-2252, 2005
4) Dürr V, Schmitz J, Cruse H: Behaviourbased modelling of hexapod locomotion: Linking biology and technical application. Arthropod Structure and Development, 33 (3), 237-250, 2004
5) El Sayed Auf A, Mösch F, Litza M: How the Six-legged Walking Machine OSCAR Handles Leg Amputations. Proceedings of the Workshop on Bio-Inspired Cooperative and Adaptive Behaviours in Robots at the SAB IX, Rome 2006
6) Rosano H, Webb B: The control of turning in real and simulated stick insects. Proceedings of the Ninth International Conference on the Simulation of Adaptive Behaviour, Lecture Notes in Artificial Intelligence volume 4095, (2006)
7) El Sayed Auf, A; Larionova, S.; Litza, M.; Mösch, F.; Jakimovski, B.; Maehle, E.: Ein Organic Computing Ansatz zur Steuerung einer sechsbeinigen Laufmaschine. AMS, 233-239, Springer-Verlag, Berlin Heidelberg 2007

Intrusion Detection via Artificial Immune System: a Performance-based Approach

Andrea Visconti, Nicoló Fusi, Hooman Tahayori

Abstract In this paper, we discuss the design and engineering of a biologically-inspired, host-based intrusion detection system to protect computer networks. To this end, we have implemented an Artificial Immune System (AIS) that mimics the behavior of the biological *adaptive immune system*. The proposed AIS, consists of a number of running artificial white blood cells, which search, recognize, store and deny anomalous requests on individual hosts. The model monitors the system through analysing the set of parameters to provide a general information on its state — ill or not. When some parameters are discovered to have anomalous values, then the artificial immune system takes a proper action. To prove the effectiveness of the suggested model, an exhaustive test on the AIS is conducted, using a server running Apache, Mysql and OpenSSH, and results are reported. Four types of attacks were tested: remote buffer overflow, Distributed Denial of Service (DDOS), port scanning, and dictionary-attack. The test proved that our definition of self/non-self system components is quite effective in protecting host-based systems.

1 Introduction

Artificial Immune Systems (AISs) are inspired by the workings of the biological immune systems [1, 2, 3], and focus their capability to recognize elementary *self*

Andrea Visconti
Universitá degli Studi di Milano, Dipartimento di Informatica e Comunicazione, Via Comelico 39/41 Milano 20135 Italy, e-mail: andrea.visconti@unimi.it

Nicoló Fusi
Universitá degli Studi di Milano, Dipartimento di Informatica e Comunicazione, Via Comelico 39/41 Milano 20135 Italy, e-mail: nicolo.fusi@studenti.unimi.it

Hooman Tahayori
Universitá degli Studi di Milano, Dipartimento di Scienze dell'Informazione, Via Comelico 39/41 Milano 20135 Italy, e-mail: hooman.tahayori@unimi.it

Please use the following format when citing this chapter:

Visconti, A., Fusi, N. and Tahayori, H., 2008, in IFIP International Federation for Information Processing, Volume 268; *Biologically-Inspired Collaborative Computing*; Mike Hinchey, Anastasia Pagnoni, Franz J. Rammig, Hartmut Schmeck; (Boston: Springer), pp. 125–135.

components of the body — endogenous or innocuous — and elementary *non-self* components of the body — exogenous or potentially pathogenic. Artificial immune systems proposed in the last decade were mainly studied and implemented for solving real-world problems (spam filtering, intrusion detection, pattern recognition, etc.). In particular, the most representative applications of artificial immune systems are from the area of computer security and fault detection [4].

Artificial immune systems, based on the recognition of self/non-self behavior, require unambiguous definitions of all permitted and/or non-permitted actions in a system. A number of ground-breaking solutions to this problem are proposed [5, 6, 7, 8, 2]. In particular, interesting approaches suggested in [5, 9] introduce the possibility of using a sequences of system calls, executed by running UNIX processes, as discriminator between normal and abnormal behavior. Moreover, references [10, 11, 12] discuss the design and testings of Lisys, a LAN traffic anomaly detector that monitors TCP SYN packets for detecting unusual requests, alerting administrator when abnormal connections are detected. In [13], authors suggested collecting data on normal and abnormal behaviors in host-based and network-based systems. Such data, collected in a realistic context, contains information that may be used for automatically detecting, analyzing and controlling future anomaly behaviors due to new and unpredictable network attacks. Tarakanov et al. in [7] described an artificial immune system based on a rigorous mathematical approach, that applied the singular value decomposition to the matrix of connection logs and mapped the users' requests into a real two-dimensional vector space. Authors argued that similar self/non-self requests lump together. In [14, 15] authors suggested interesting approaches based on the Danger Theory [16]. This new theory has shifted control of immunity to the tissues that need protection. Inspired by the behavior of innate immune system, Pagnoni and Visconti in [13] illustrated an artificial immune system based on the working of the macrophages. The authors argued that the main idea of an intrusion detection system is not to recognize and kill a specific intruder in the most effective way, but rather to find and kill any intruder as soon as possible. In addition to the artificial immune systems previous mentioned, authors in [5, 17, 8, 18] suggested applying the negative selection algorithm to the problem of network intrusion detection.

Unfortunately, none of these solutions has achieved one hundred percent precision. Nevertheless, real-world applications have the necessity of (a) providing a strong, reliable discrimination between normal and abnormal behavior and (b) maintaining a complete database of "good or bad behavior" to be used by the self/non-self recognition algorithms. The choice of self/non-self behaviors is crucial because some bad behavior not stored in the database may not be recognized as network attacks; moreover a large database of self/non-self behaviors entails a substantial degree of slowness that is not acceptable in real-time applications.

In order to overcome these problems, we have designed a biologically-inspired intrusion detection system based on the paradigms of the *acquired immune system.* Being more slow than the innate one, the acquired immune system is the only one that remembers the previously encountered attacks, recognizes new attacks of unwanted intruders entering the system, and provides a proper response to the attack of the enemies. Diagnosing an abnormal behavior of a specific type requires knowing which, if any, set of parameters characterizes the anomaly. This set of parameters is called *antigen signature*. Some such signatures are well-known, and can be easily recognized automatically, others are just less well-defined, and can be more difficult to recognize, whilst others are completely unknown. To this end, we analyze and improve existing solutions in computer security through the design, engineering and testing an intrusion detection system that recognizes anomalous values of parameters of a given system. These anomalous values can be interpreted as a sign of an attack to the system comparable with fever in the case of presence of infection in body.

In this paper, we suggest an artificial immune system based on several agents that mimics the behavior of white blood cells in the acquired immune system. Such white blood cells — or lymphocytes, — cooperate using a specific communication protocol in order to protect the system against exogenous or endogenous attacks. Every white blood cell is a separate process that monitors the parameters of the system and checks the presence of non-self antigen signatures.

In the sequel, in section 2 we discuss the principles of immune systems, in particular, we focus our attention on the adaptive immune systems. In Section 3, we describe the design and engineering of our artificial immune system; while an intensive testing is presented in section 4. Finally in Section 5, we provide pros and cons of our model and draw conclusions.

2 Biological Immune System

Biological immune systems draw up several lines of defense to protect the organism. This defense systems include *chemical* and *physical barriers*, *innate immune system* and *adaptive immune system*.

Chemical and physical barriers provide the first line of defense in the fight against invaders. Examples of the chemical and physical barriers are skin, gastric acid in the stomach, eyelashes, tears, and so on. These barriers try to protect the body against pathogens that enter an organism, and consequently reduce the probability that the pathogens will lead to an illness. Unfortunately, some foreign invaders that are present on the skin surface pass through injuries on the skin.

When the chemical and physical barriers fail to stop unwanted intruders, invaders are attacked by the cells of the second line of defense: the *innate immune system*. The innate immune system recognizes and attacks invaders in a generic way, with no necessity of previous exposure to them. The cells involved in the innate reaction are *leukocytes* such as macrophages, natural killer cells, mast cells, basophils, and so on. These leucocytes (a) release chemical factors that cause inflammation, swelling and local blood vessel dilation; (b) recruit immune cells to sites of infection; (c) attack everything of a foreign nature, engulfing pathogens and dead cells in a process called phagocytosis; (d) and finally, activate the adaptive immune system.

The *adaptive* or *acquired immune system* provides the third line of defense in the fight against intruders. It does not replace the innate immune system, but rather improves it. The ability of the adaptive immune system to kill invaders is based on the capacity of recognizing several kinds of pathogens and remembering specific antigen signatures after the resolution of the infection. Comparing to the previous two lines of defense, the adaptive immune system works in a more complex way because its responds to an attack is antigen-specific. Being exposed to different pathogens, the adaptive immune system learns to identify enemies and as a result its specific response will be more effective than a generic response of the innate immune system.

The cells involved in the acquired reaction are *leukocytes* such as memory B cell, killer T cell, helper T cells, and so on. When activated, these leucocytes are able to (a) distinguish the cells of the body from unwanted invaders; (b) recognize specific signature for each non-self antigen; (c) generate a specific immune response against invaders; (d) remember specific signatures for each non-self antigen; (e) and eventually, quickly remove the previously encountered non-self antigen.

Unfortunately, these lines of defense, that generate a powerful barrier against intruders, are not perfect and sometimes fail. Failures occur when the ability to fight invaders of one or more components is reduced — immunodeficiency — or the ability to recognize self and non-self cells is compromised — autoimmunity. In both cases, an organism will be vulnerable to infections.

3 Design and Engineering of Artificial Immune System

To achieve the ultimate goal of designing and engineering an intrusion detection system (IDS) based on the workings of the acquired immune system, clear understanding of the characteristics of the acquired immunity is of great importance. In particular, we concentrated on (a) the acquisition of a clear discrimination between self — regular — and non-self — unwanted — system behaviors, (b) the elimination of recognized infections — recognized attacks —, (c) the take care of system injuries — bugs, — (d) the detection and elimination of new infections — new attacks, — and (e) the absence of autoimmune reactions.

The design and engineering of the proposed AIS is based on the previous points and on the observation that getting into a system without leaving any track is virtually impossible. We search for these tracks by considering the parameters of the system when our server is under attack. In order to identify such tracks i.e. the fingerprint of attacks, the values of different system parameters were surveyed. The gathered data was analyzed and more than 150 graphs were generated. Some of them did not show any significant change during an attack, while some did. We conceived the artificial immune system with these ideas in mind.

The proposed AIS is a host-based system that consists of a set of processes — helper T-cells, killer T-cells, and memory B-cells — running on a server. These processes act and cooperate as digital lymphocytes in order to discover suspicious values of the parameter of the system and face external attacks. The artificial immune system must be initialized through a training phase, in which the AIS defines the number of running lymphocytes. In fact, this number is not constant, but is optimized experimentally because it depends on the hardware features and the workload of the server. This number can vary between a lower and an upper bound, improving the performance of the AIS when the system is under attack. Furthermore, if such number falls under the optimized threshold, new lymphocytes are automatically created. In addition to the task of defining the number of running lymphocytes, the training phase is also responsible for setting up the parameters of all processes. Indeed, the digital lymphocytes learn to identify self/non-self behaviors, analyzing the data of system parameters under the supervision of an expert. After the training phase, the artificial immune system is ready to be activated.

Helper T-cells Are processes, or agents, of the artificial immune system. Their main objectives are identifying an anomalous behavior in the monitored parameters and promoting the activation of adaptive immune response. In order to do so, helper T-cells collect a sample data of actual system parameters, compute the mean and the standard deviation of such data, and compare the current values to the previously stored values. If the mean of each parameter monitored exceeds a given threshold — mean stored in memory plus or minus the standard deviation stored in memory, — the current interval is defined non-self. Moreover, helper T-cells try to identify the type of attack using type-1 fuzzy rules and promoting a quick immune response. Indeed, they stimulate other cells of immune system, controlling and inhibiting immune attacks against self-antigens.

Every helper T-cell has a lifespan at the end of which the cell dies. This means that the cell will be regenerated, will undergo the negative selection phase, and hence the ability to recognize possible unwanted attacks will be improved.

Memory B-Cells Are processes that remember attacks previously encountered. When they recognize a set of parameters that show an anomalous behavior previously identified, the artificial immune system stops the recognition phase and memory B-cells inform killer T-cells that the system is under attack, specifying what kind of attacks is.

Killer T-cells Are processes that take proper actions for denying and eliminating unwanted behaviors. If the attack type is known, a predefined action will take place. For example, in presence of a denial of service attack, killer T-cells deny unwanted requests, banning the IP addresses of the senders, while in presence of a reverse shell techniques, killer T-cells eliminate such shells. On the other end, if the attack type is unknown, a notification is sent to the system administrator, alerting him of the security threat.

All digital cells are separate Java processes, so if one process crashes it will not affect the working of the artificial immune system. Designing the system as a set of different processes offered greater security and stability, at the price of a more difficult communication between processes. In order to solve this problem, we implemented a communication protocol. When authenticated, processes can communicate with each other by calling specific functions. This communication protocol is necessary to stimulate groups of lymphocytes, or the entire artificial immune system, when the system is under attack.

4 Testing

All tests were made on a dual Intel® PIII® server with 1.5Gb of RAM on which Gentoo linux [19] was installed; running Apache 2.0.59 with PHP 5.2.5, Mysql 5.0.40 and OpenSSH 4.7. Several types of attacks were tested: port scanning (NMAP), remote buffer overflow, Distributed Denial of Service (DDOS) and dictionary-attack against SSH authentication. These types of attack were chosen because they have very different performance fingerprints and are the most common.

In order to simulate a real-world situation, a fake institutional website was implemented and the system was tested with different amounts of traffic:

None   No one, or a small number of users, surf the website. This profile has been used in order to provide baseline data.

Moderate   An average number of users surf the website.

Intense   A really high amount of users surf the website, making a huge number of self requests. This profile has been used in order to evaluate the behavior of the system under severe stress condition.

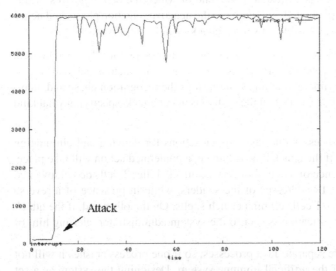

Figure 1 Interrupts per second during a DDOS attack without legitimate traffic

The population of digital cells used for the testing is as follows: 3 to 5 killer T-cells, 5 to 10 helper T-cells and 3 to 5 memory B cells.

The traffic has been simulated with JMeter [20], a stress testing tool for web applications provided by the Apache software foundation.

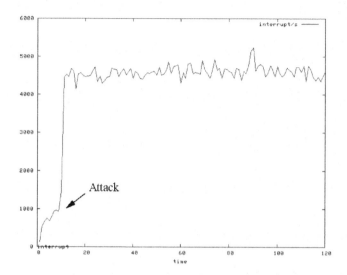

Figure 2 Interrupts per second during a DDOS attack under moderate traffic

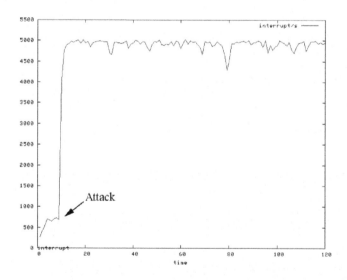

Figure 3 Interrupts per second during a DDOS attack under intense traffic

As mentioned in Section 3, our AIS monitors the system parameters in order to identify unusual patterns that may be related to unwanted behaviors. For example, figures 1, 2 and 3 show the behavior of a system parameter during a DDOS attack. Analyzing the values of this parameter, an unusual pattern in the number of interrupt requests can be recognized.

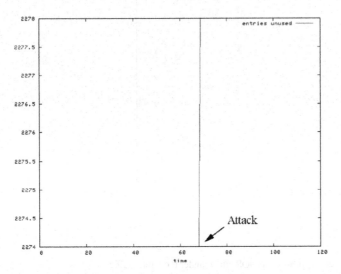

Figure 4 Unused cache entries during a Buffer Overflow attack under moderate traffic

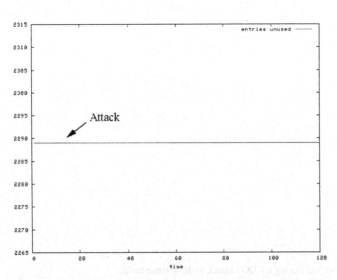

Figure 5 Unused cache entries during a Buffer Overflow attack under intense traffic

Figures 1, 2, and 3 show clearly that DDOS attack is rather easy to spot, because it largely affects the system performance. Unfortunately, as can be seen in the figures 4 and 5, not all kinds of attack are always so easy to recognize. Such figures represent the unused cache entries during a buffer overflow attack while an average and large number of users, respectively, surf the website. In the first two cases, — figures 4 and 5, low and moderate workload — the artificial immune system is able to recognize the anomaly. In the last case — figure 6, high workload — the artificial immune system fails. It is easy to see that a high workload situation may introduce an excessive level of background noise, decreasing the ability of the system to recognize anomalous behaviors. Such situations affect negatively the performance of the artificial immune system, enhancing the risk of false positives.

Tables 1, 2, and 3 summarize the results of the testing activity. For each of the four attacks tested, we mark the set of system parameters that may indicate the presence of an attack.

Table 1 No traffic

Attacks	UsedIH	UnusedCE	PGfault	SysCPU	Interr	TRate
Buf Overfl.	X	X	X			X
DDOS				X	X	
Scan NMAP	X	X	X	X		X
Bruteforce			X	X		

Table 2 Moderate traffic

Attack	UsedIH	UnusedCE	PGfault	SysCPU	Interr	TRate
Buf Overfl.	X	X	X			
DDOS				X	X	
Scan NMAP						
Bruteforce			X	X		

Table 3 Intense traffic

Attack	UsedIH	UnusedCE	PGfault	SysCPU	Interr	TRate
Buf Overfl.	X		X			
DDOS				X	X	
Scan NMAP	X	X				X
Bruteforce			X	X		

As tables 1, 2, and 3 illustrate, recognizing an attack in the presence of many net-surfing users is increasingly difficult, and in some cases is rather impossible.

For example, recognizing a port scanning attack is almost impossible, given the anomalous behavior of the monitored parameters under different traffic profiles (see tables 1, 2 and 3).

The results of an exhaustive testing are summarized in table 4.

Table 4 Test results

	No	Moderate	Intense
Buf Overfl.	95%-100%	75%-90%	65%-85%
DDOS	100%	95%-100%	90%-100%
Scan NMAP	0%-20%	0%	0%-5%
Bruteforce	100%	90%-100%	85%-100%

5 Conclusions and Future Works

The suggested artificial immune system is an IDS based on the idea of equipping servers with the technological equivalence of an acquired immune system. To this end, our AIS monitors and analyzes a set of system parameters to check for anomalous behaviors. Although still at a preliminary stage, the exhaustive testing revealed that AIS is able to quickly detect anomalous behaviors previously encountered, deny proliferation of foreign processes by killing dangerous processes before they will widely used, and recognize attacks with a strong fingerprint such as denial of service and dictionary attack. On the other hand, the intensive testing has proved that in order to avoid a large number of false positives, we have to lower the sensibility of the system, affecting the recognition of some kinds of attack. Indeed, under these circumstances the system cannot recognize an NMAP scan; moreover, in presence of many net-surfing users recognition is increasingly difficult, and sometimes is impossible.

It should be stressed that yet, no single method, biological or artificial, can achieve one hundred percent precision. For these reasons, the suggested system is not meant to replace firewalls, login policies, or antivirus because it cannot blocks every kind of attack. The proposed artificial immune system should be used in conjunction with other complementing technologies either biologically inspired or not.

Our future works will be devoted to improve the AIS, extending actual acquired immunity with specific components of a second line of defense: the innate immunity.

6 Acknowledgements

This research was funded by the State of Italy via FIRST (Fondo per gli Investimenti nella Ricerca Scientifica e Tecnologica).

References

1. D'haeseleer, P., Forrest, S., Helman, P.: An immunological approach to change detection: algorithms, analysis and implication. In: Proceedings of the 1996 IEEE Symposium on Computer Security and Privacy, (1996)
2. Forrest, S., Hofmeyr, S., Somayaji, A., Longstaff T.: A sense of self for UNIX processes. In: Proceedings of the 1996 IEEE Symposium on Research in Security and Privacy, (1996)
3. Forrest, S., Hofmeyr, S., Somayaji, A.: Computer immunology. In: Communication of ACM **40**(10), 88-96 (1997)
4. Dasgupta, D.: Advances in Artificial Immune Systems. In: IEEE Computational Intelligence Magazine, (November 2006)
5. Hofmeyr, S., Somayaji, A., Forrest, S.: Intrusion Detection using Sequences of System Calls. In: Journal of Computer Security **6**(3), 151-180 (1998)
6. Dasgupta, D.: Immune-based intrusion detection system: A general framework. In: Proceedings of the 22nd National Information Systems Security Conference, (1999)
7. Tarakanov, A.O., Skormin, V.A., Sokolova, S.P.: Immunocomputing: Principles and Applications. Springer-Verlag, New York (2003)
8. Forrest, S., Glickman, M. R.: Revisiting LISYS: Parameters and Normal behavior. In: Proceedings of the 2002 Congress on Evolutionary Computation, (2002)
9. Warrender, C., Forrest, S., Pearlmutter, B.: Detecting intrusions using system calls: Alternative data models 1999. In: IEEE Symposium on security and Privacy, (1999)
10. Hofmeyr, S., Forrest, S.: Architecture for an artificial immune system. In: Evolutionary Computation, **8**(4), 443-473 (2000)
11. Hofmeyr, S.: An immunological model of distributed detection and its application to computer security. In: PhD thesis, University of New Mexico, (1999)
12. Balthrop, J., Forrest, S., Glickman, M.: Revisiting lisys: Parameters and normal behavior. In: Proceedings of the Congress on Evolutionary Computation, (2002)
13. Pagnoni, A., Visconti, A.: An Innate Immune System for the Protection of Computer Networks. In: Proceedings of the 4th International Symposium on Information and Communication Technologies, (2005)
14. Aickelin, U., Cayzer, S.: The Danger Theory and Its Application to Artificial Immune Systems. In: Proceedings of 1st International Conference on Artificial Immune Systems, (2002)
15. Aickelin, U., Bentley, P., Cayzer, S., Kim, J., McLeod, J.: Danger Theory: The Link between AIS and IDS? LCNS 2787, (2003).
16. Anderson, C., Matzinger, P.: Danger: the view from the bottom of the cliff. In: Seminars in Immunology, **12**(3), 231-238 (2000)
17. Kim, J., Bentley, P.: The human Immune system and Network Intrusion Detection. In: Proceedings of 7th European Congress on Intelligent techniquesSoft Computing, (1999)
18. Gonzalez, F., Dasgupta, D.: An Immunogenetic Technique to Detect Anomalies in Network Traffic. In: Proceedings of the International Conference Genetic and Evolutionary Computation (GECCO), (2002)
19. Gentoo linux, available at *http://www.gentoo.org/*
20. Apache JMeter, available at *http://jakarta.apache.org/jmeter/*

Immuno-repairing of FPGA designs

Norma Montealegre, Franz J. Rammig

Abstract FPGAs can be used for the design of autonomic reliable systems. Advantages are reconfiguration and flexibility in the design. However commercial FPGAs are first prone to errors. Second, the design flow is not yet supported for the use of fault tolerance techniques like Built-In Self-Tests. Fault tolerance can be reached through error detection and fault recovery. Most error detection techniques are not suitable for on-line detection because of detection times and long and inflexible training. This paper proposes a fault tolerant design for FPGAs. It has a Built-In Self-Test which error evaluation and fault recovery is supported by computing techniques inspired in the Immune System. A fault recovery and a hardware implementation model are also to be presented.

Keywords: autonomic systems, fault tolerance, immunocomputing, FPGA, BIST.

1 Introduction

Nowadays there is the demanding requirement of having systems which faults can be recovered without human intervention. That is the field of autonomic reliable systems. Autonomous robots and vehicles in outer space and undersea systems [8] are prone to errors due to its dynamic and environment of action. These systems are designed with radiation-hardened or higher and lower temperature range components, like radiation-hardened FPGAs [10]. Hardware design techniques based on Triple Modular Redundancy help in developing FPGA-based circuits resilient to SEUs (Single Event Upset) [7], like the tool referred in [11] or the TMR-Tool from Xilinx [21]. While circuits are hardened with special components and TMR has a limited fault recovery, a seamless design flow for fault recovery is not present yet.

Some algorithms were developed in the field of Artificial Immune Systems, inspired in the vertebrate's immune system. They have served in solving computing problems. Nevertheless those algorithms have inspired also electronic designs in the

Norma Montealegre
Heinz Nixdorf Institute, Fuerstenalle 11, 33102 Paderborn, Germany
e-mail: norma@upb.de

Please use the following format when citing this chapter:

Montealegre, N. and Rammig, F.J., 2008, in IFIP International Federation for Information Processing, Volume 268; *Biologically-Inspired Collaborative Computing*; Mike Hinchey, Anastasia Pagnoni, Franz J. Rammig, Hartmut Schmeck; (Boston: Springer), pp. 137–149.

searching of fault tolerance. One example is the research coined as "Immunotronics" (Immune + Electronics), the hardware fault tolerance inspired by the immune system [3]. In [18] a centralized immune layer is presented. The learning phase identifies correct operation of the systems as "self" building antibody patterns composed of: inputs, excitation, correct states and outputs. In the operational phase, the identification of "non self" operations is implemented by means of genetic algorithms. On the other side, a decentralized immune layer has taken inspiration from cell biology to create a multicellular FPGA. This idea emerged the field of Embryonics which together with Immunotronics generate a two level structure [2]. The first level is composed of the embryonic cells which communicate across data channels. A second layer is composed by antibodies which communicate also across data channels and are named together the lymphatic network. Trans-layer communication channels are present between antibodies and embryonic cells as well. Antibodies store self-tolerance conditions. Furthermore, healing of the embryonic cells is regarded by [17], who considers a cell's self-test for the fault diagnosis, and a cell repair or elimination through reconfiguration of the cell's routing. All these methods carry to a new hardware conception not available in commercial FPGAs.

Immunocomputing explores, in a formal way, the principles of information processing that proteins and immune networks utilize in order to solve specific complex problems [16]. Free binding among proteins inspired Formal Immune Networks, which are able to learn, recognize and solve problems. This method is based in the Singular Value Decomposition of a matrix. It proved to have small learning and recognition times and a good resource efficiency regarding memory and computing [13]. Moreover, it presents a self-organizing property since, for a training set, iterations within the algorithm self-converge to antibodies [13]. Making use of its efficiency, Immunocomputing can be used for on-line error detection [15].

A self-test system has two main components: a test pattern generator and a test response evaluator. The test patterns and expected responses are stored in a memory. It is necessary a control signal to turn on the testing, a counter to address the memory and a comparator for comparing the obtained response with the expected one [9]. Because of the quantity of test patterns, this approach is time consuming. There are some alternatives of output response analysis in which output data compaction takes place. One of them are concentrators, counting techniques, signature analysis, accumulators, comparators, etc [12]. Comparison-based response evaluators compare on a vector-by-vector basis the expected responses stored in a memory and the output responses of the circuit under test. This approach is simple and modular. Besides a distributed Built-In Self-Test with n-test pattern generators, n-circuits under test and one test response evaluator can be applied. A BIST system can also work on-line [1]. The potential problem is the long time that may be required to cycle through the test patterns and evaluate the responses before determining if an error is present or not, [5] approaches this problem. This is critical for systems where the BIST works on-line and fault recovery should be done at time. A molecular approach is given by [4] and [19], but a circuit oriented design is not taken into account. Therefore a correct partitioning of the circuit, a distributed BIST with a fast response evaluator and fault recovery support is needed. This paper is a contribution following this tendency.

Fault tolerance can be reached through error detection and fault recovery. The present paper proposes a distributed fault tolerant design for FPGAs using Formal Immune Networks and self-test systems. The system has a distributed error detection mechanism through distributed Built-In Self-Tests inside a FPGA, see Fig. 1. BIST synthesis for a very large design may be possible within linear time by extracting sub-circuits which are almost constant in size [6]. That accelerate logic BIST synthesis procedures and reduces the time error detection takes. The circuit under test is one part of a partitioned circuit. The circuit receives a test pattern and the response is evaluated by means of cFINs (cytokine Formal Immune Networks) [14]. BIST can profit of the celerity offered by the cFIN method in detecting errors applying a determined error correction method at the proper time. Test response evaluation and fault recovery by cFIN for fault tolerant FPGA circuit designs is the main contribution of the present work. The decentralized BIST procedure is controlled by a global test scheduler module, a fault processing mechanism and a fault recovery module. A hardware implementation of the whole system is also proposed.

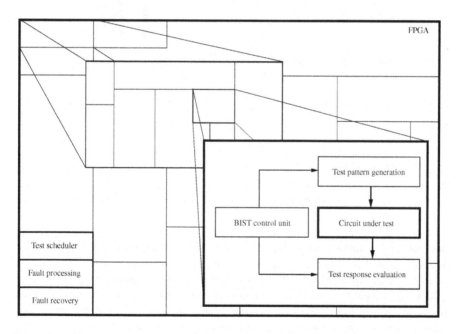

Figure 1 *Diagram of the proposed system*

2 Built-In Self-Test Proposal

Figure 2 shows a BIST proposal with on-line learning. The system needs to be trained with the test patterns before being operational. For test response evaluation purposes, outputs of the Circuit Under Test are evaluated with a method inspired in the cytokine-Formal Immune Networks and presented in Section 5. Making a biological analogy, an *antibody* represents the *expected output* transformed into the Formal Immune Network space. An *antigen* is the *response of the circuit under test*. A *cytokine* represents the *action* to be taken for fault recovery purposes.

It is important to note that after training of the system, on-line learning can take place. This is possible mapping the value of the new A_i test pattern into the cFIN space. This point is added to the compacted expected response data (compaction or compression performed by means of cFIN). Therefore, in case of a change in the training patterns, the training phase does not necessarily have to be repeated with the entire training set [13].

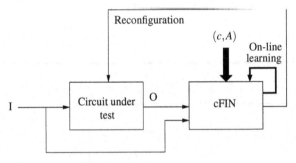

Figure 2 *Built-in-self-test with a cFIN on-line learning*

If this BIST model is applied for a whole circuit, the complexity in building the training matrix and the time for training and recognition may explode. Therefore, circuit partitioning [6] is considered, see Fig.1. Methods for circuit partitioning are not the scope of this paper.

A training matrix $V(c,A)$ should be provided prior to the operation of the system (test pattern generation). A is a matrix with information over expected responses under defined inputs. Each expected response should be linked to a recovery procedure in case of failure, expressed by c. In case of combinational circuits, training patterns are composed of Input/Outputs. But, sequential circuits consider also stimuli and internal states, as seen in Fig. 3. For test purposes, such circuits may be transformed to a sequence of combinational ones using conventional scan-path techniques.

The training matrix should regard the procedure for fault recovery under failure. For every training pattern is recommended to have a recovery alternative expressed in an integer coded value c, see Fig. 4. c represents a cytokine that signals the action to be taken at the time of finding an error.

The BIST Control Unit supports the hierarchical BIST strategy shown in Fig. 1. It contains an input for starting the BIST and an output for indicating the end of the test. A pattern counter determines the ending of a test. Two schemes to be considered are possible at the time of designing the BIST, the *test-per-scan* and the *test-per-clock*

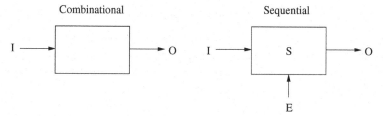

Figure 3 *Consideration at the moment of building the training matrix*

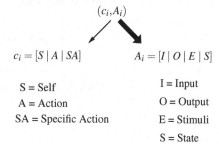

$$c_i = [S \mid A \mid SA] \qquad\qquad A_i = [I \mid O \mid E \mid S]$$

S = Self	I = Input
A = Action	O = Output
SA = Specific Action	E = Stimuli
	S = State

Figure 4 *Test cases*

scheme. It is not the aim of this paper to give a detailed description. Please refer to [20].

3 Global BIST

In order to apply a consistent error correction, the test schedule, fault processing and fault recovery are global modules for the whole system.

Depending on the application, tests can be recurrent or preemptive. In the case of a preemptive one, time error recognition should be considered in order to plan the frequency of testing. Frequency of testing is a function of the clock rate as well. A *test session* is a set of test unit processed in parallel and a *BIST schedule* is a series of test sessions which is implemented by the BIST Control Unit in hardware.

The class or cytokine's natural value represents the action to be taken in the design when a failure occurs. It has to be specified which recovery method should be applied for a specific failure. This data should be provided together with the training matrix in the case of *Supervised Learning* but it could also be determined after the training process by clustering points in the mapped FIN space. That is the case of *Unsupervised Learning*. In the first case this array can be constructed with the following data:

$$c = \{S \mid A \mid SA\} \tag{1}$$

Where:

S   *Self* considers whether the recognized pattern is a failure "non-self" or a particular pattern. This can be used not only for failure detection, but also for warning states not considered as malicious.

A   *Action* considers a general action to be taken i.e. total reconfiguration.

SA   *Specific Action* considers a more refined method to recover from the failure.

Fault recovery is based on the reconfigurabilty property of FPGAs. Therefore, the failure recovery can be executed by total or partial reconfiguration. Other alternatives for fault recovery are application dependent and should be addressed at the time building the EA array.

4 Hardware Implementation

The proposed BIST can be implemented as an Intellectual Property Core inserted into the same FPGA as the circuit to be tested, Fig. 5. In this case, faults present in the Circuit Under Test are also prone to appear in the BIST. An alternative is to provide an external second FPGA which implements only this procedure.

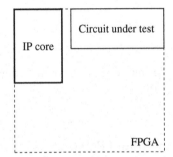

Figure 5 *Implementations as*
an IP core

It is also possible to consider an external circuit composed of a DSP and a micro-controller, like the one in Fig. 6. The DSP is able to compute in parallel the mapping of points to the FIN and to compare distances among points [13]. The micro-controller could implement modules of the global BIST. Nevertheless, faults in the connection path between the circuit under test and the self-test system should be regarded in this implementation case.

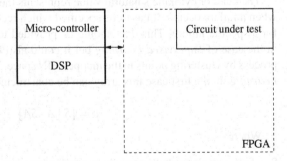

Figure 6 *Implementation as*
an external circuit

5 c-Formal Immune Networks

This section explains the method of training and recognition of a cFIN. This method is used in the implementation of the response evaluation and fault processing of the Built-In Self-Testing system. For a more detailed and extensive explanation of this theory, please refer to [14], [13] and [15].

Immunocomputing intends to establish a new kind of computing. The main difference with other kinds of computing lays in its basic element, the *formal protein*. A protein is an essential component of organisms and participate in every process within cells. Proteins constitute epitopes present in antigens and antigen presenting cells. Proteins constitute also paratopes present in antibodies. Epitope is the minimum molecular structure that is able to be recognized by the immune system. One epitope matches with a paratope in molecular recognition. Figure 7 shows the antigen binding site of an antibody named as *paratope* that recognizes the epitope of an antigen or an antigen presenting cell. An antigen presenting cell is a cell that has digested an antigen and presents in its surface an epitope. An epitope is made of around 10 amino-acids. The same applies to a paratope. A protein is composed of amino-acids arranged in a linear chain. The 3D shape or tertiary structure of the epitope is recognized by a paratope, see Fig. 7. It means, an epitope is a kind of surface protein. That is why proteins will be seen as the basic element in Immunocomputing.

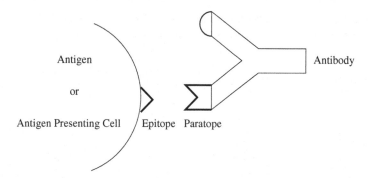

Figure 7 *The epitope of one antigen or an antigen presenting cell is recognized by the paratope of an antibody*

Cytokines are also introduced. Cytokines are groups of proteins secreted by many types of cells. Each cytokine binds to a specific cell's surface receptor signaling a specific action i.e. differentiation into plasma cells, antibody secretion or cell death. They bind also through own receptors constituted from proteins too, see Fig. 8.

B-cells in the immune system secrete antibodies. They also secrete cytokines in order to signal something to another cell. Then, a B-cell will be taken as a generic cell V_i with two components expressed by:

$$V_i = (c_i, P_i) \tag{2}$$

Where:

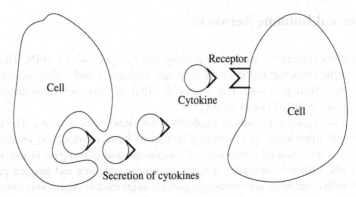

Figure 8 *Cytokines signal the cellular interaction. They are secreted by cells. They are recognized by cell's receptors*

$c_i \in \mathbb{N}$ represents a cytokine. Recovery action to be taken under presence of error. $P_i \in \mathbb{R}^q = ((p_1)_i, ..., (p_q)_i)$ is a point in a q-dimensional space. P lies within a cube $max\{|\,(p_1)_i\,|, ..., |\,(p_q)_i\,|\} \leq 1$. It represents a protein transformed into the FIN (Formal Immune Network) space. In biological terms it represents an antigen binding site of an antibody or simplifying an antibody. An array containing an input test and its test response, all transformed to the FIN space is an antibody.

In Fig. 9, a two dimensional Formal Immune Network (2D-FIN) is presented. As $q = 2$, each protein has two coordinates in the FIN space.

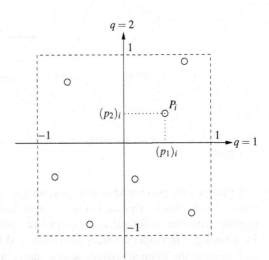

Figure 9 *2D-FIN. Note that q=2 and P represents an antibody (protein) in the FIN space*

5.1 First stage training

Training consist in transforming a given energy matrix A into another antibody matrix P. Matrix A is a composed of arrays of test inputs and its corresponding test response outputs, see Fig. 4. Training is the process of mapping antibodies into the Formal Immune Network space. n dimension training patterns are transformed to reduced dimension patterns (two or three). This takes place by means of Singular Value Decomposition, restricting its terms of decomposition to two or three. Figure 10 introduces the general concept. Singular values and the right singular vectors are used for the calculation of the coordinates of each training pattern into the FIN space. Each point will represent an antibody. In the figure, two outputs are displayed. First, the SVD: singular values, right singular vectors and left singular vectors. Second, the matrix P with the transformed patterns, where each pattern remains linked to its initial c value. The output should be stored in order to be used in the later stages.

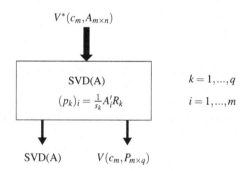

$$V^*(c_m, A_{m \times n})$$

$$SVD(A) \qquad k = 1, ..., q$$
$$(p_k)_i = \frac{1}{s_k} A_i' R_k \qquad i = 1, ..., m$$

$$SVD(A) \qquad V(c_m, P_{m \times q})$$

Figure 10 Training

A can be written as a linear combination of pairwise orthogonal projections:

$$A = s_1 L_1 R_1' + s_2 L_2 R_2' + s_3 L_3 R_3' + ... + s_k L_k R_k' + ... + s_r L_r R_r' \qquad (3)$$

where:

r   rank of the matrix A
L_k left singular vectors
R_k right singular vectors
s_k singular values

Moreover, L_k is of dimension m and R_k of dimension r.

The minimal binding energy is achieved with the pair of proteins whose angles in its spacial configuration form singular vectors. Those singular vectors correspond to the maximal singular values of the matrix A. Singular vectors represent formal protein probes and the singular values their binding energy. As the singular values are ordered in a decreasing order, we can take the first two singular values and its corresponding terms for a 2D-FIN and three terms for a 3D-FIN. In consequence every training vector pattern of dimension n is mapped to only two (or three) values of binding energy in the FIN space [16]. Afterwards, it is necessary to map the training vectors into the FIN space (see Fig. 9) by means of:

$$(p_k)_i = \frac{1}{s_k} A'_i R_k \qquad (4)$$

where $i = 1, ..., m$ and $k = 1, ..., q$.

5.2 Second stage training

The *affinity* between two cells V_i and V_j, can be calculated by the distance d_{ij} given by:

$$d_{ij} = max\{| (p_1)_i - (p_1)_j |, ..., | (p_q)_i - (p_q)_j |\} \qquad (5)$$

Initially given m cells with pairs cytokine-antibody $V_i(c_i, P_i)$, the aim is to reduce similar cells killing one of two cells which distance is less than a given *threshold*. The set of m cells belonging to the *innate immunity* can be represented by a W_0 as shown below:

$$W_0 = \{V_1, ..., V_m\} \qquad (6)$$

cFIN means cytokine Formal Immune Network. A cFIN is a set of cells $W \subseteq W_0$. In contrast with a FIN, it considers a second stage training or *maturation*, inspired in the cytokines from the immune system. [14] introduces a two stage second training in order to get a reduced set of cells, therefore improving the resource utilization and the time applied in recognition in the future.

The first stage is *apoptosis* and it intends to reduce the set with the following rule:

If $V_i \in W$ recognizes $V_k \in W$, then remove V_k from cFIN.

Note that *recognition* means:

$$c_i = c_k \qquad (7)$$
$$d_{ik} \leq h \qquad (8)$$

Where h is a threshold of affinity.

The second stage is *auto-immunization*. It tries to recover accidentally removed cells by the process of apoptosis.

The removed cell V_i nearest to a cell V_k from the set W will be inserted again if $c_i \neq c_k$.

Figure 11 shows graphically the sets and the general concept of the optimization offered by cFIN.

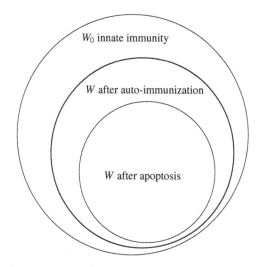

Figure 11 *Sets of cells after Apoptosis and Auto-immunization*

5.3 Recognition

Figure 12 shows an antibody close to an antigen. When the distance between any antibody and the antigen is less than a given *threshold*, recognition in the FIN space is produced.

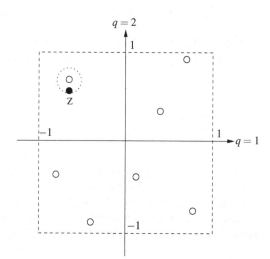

Figure 12 *An antibody recognizes an antigen Z inside its affinity threshold radio*

An antigen $Z = [z_1, z_2, ..., z_n]$ can be seen as an epitope, therefore as a protein, see Fig. 7. An antigen represents the test response linked with its test input. In order to be compared with the antibodies, it should be mapped to a point in the q-dimensional FIN space by:

$$p_k = \frac{1}{s_k} Z' R_k \qquad (9)$$

p_k values should be mapped into the FIN space. See also Fig. 13.

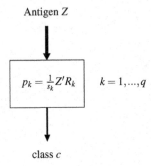

Figure 13 *Recognition*   class c

> Cell V_i recognizes antigen Z by assigning it a class c_i, if the distance between the antigen among all antibodies of the cFIN is $d(V_i, P) = min\{d(V_j, P)\}$, for all $V_j \in W$. A test response will be matched with the expected output recognizing whether there is an error or not and applying the determined action signaled by c_i.

For the distance calculation Eq. 5, the Euclidean norm is taken. Nevertheless, the choice of the norm is determined by the appearance of the group of points in the FIN space.

Relating the class c in Fig. 13, it is a numerical value which can also be taken as a symbolic value like "good", "bad", "reconfiguration", etc.

6 Conclusion

Computation times for the training and recognition presented in [13], show that it is feasible to expect a good performance of the model in hardware. Furthermore, the reduced memory constraints obtained with the second training of the cFIN indicates potential towards a distributed error detection and correction scheme. This paper is intended to present a design idea. Currently, an implementation with commercial components and the measure of performance is being carried out.

References

1. Al-Asaad, H., Shringi, M.: On-line built-in self-test for operational faults. In: AUTOTESTCON Proceedings, 2000 IEEE, pp. 168–174. Springer-Verlag New York (2000)
2. Bradley, D., Ortega-Sanchez, C., Tyrrell, A.: Embryonics + immunotronics: A bio-inspired approach to fault tolerance. In: J. Lohn, A. Stoica, D. Keymeulen (eds.) The Second NASA/DoD workshop on Evolvable Hardware, pp. 205–224. IEEE Computer Society, Palo Alto, California (2000)

3. Bradley, D.W., Tyrrell, A.M.: Immunotronics: Hardware fault tolerance inspired by the immune system. In: ICES, pp. 11–20 (2000)
4. Dutt, S., Verma, V., Suthar, V.: Built-in-self-test of fpgas with provable diagnosabilities and high diagnostic coverage with application to online testing. In: Computer-Aided Design of Integrated Circuits and Systems, IEEE Transactions on, vol. 27, pp. 309–326. IEEE (2008)
5. Gupta, S.K., Pradhan, D.K.: Utilization of on-line (concurrent) checkers during built-in self-test and vice versa. In: IEEE Transactions on Computers, vol. 45, pp. 63–73. IEEE Computer Society Washington, DC, USA (1996)
6. Irion, A., Kiefer, G., Vranken, H., Wunderlich, H.J.: Circuit partitioning for efficient logic bist synthesis. In: Design, Automation and Test in Europe, 2001. Conference and Exhibition 2001. Proceedings, pp. 86–91 (2001)
7. Kastensmidt, F.L., Carro, L., Reis, R.: Fault-Tolerance Techniques for SRAM-Based FPGAs, *Frontiers in Electronic Testing*, vol. 32. Springer (2006)
8. Kumagai, J.: Swimming to europa. IEEE Spectrum **44**(9) (2007)
9. Rammig, F.J.: Systematischer Entwurf digitaler Systeme. B. G. Teubner Sttutgart (1989)
10. Ratter, D.: Fpgas on mars. XCell journal **50** (2004)
11. Sterpone, L., Violante, M.: A design flow for protecting fpga-based systems against single event upsets. In: Defect and Fault Tolerance in VLSI Systems, 2005. DFT 2005. 20th IEEE International Symposium on, pp. 436–444 (2005)
12. Stroud, C.E.: A Designer's Guide to Built-in Self-Test. Springer (2002)
13. Tarakanov, A.O.: Formal Immune Networks: Self-Organization and Real-World Applications, *Advanced Information and Knowledge Processing*, vol. Part III, pp. 271–290. Springer London (2008)
14. Tarakanov, A.O., Goncharova, L.B., Tarakanov, O.A.: A cytokine formal immune network. In: Advances in Artificial Life, *Lecture Notes in Computer Science*, vol. 3630. Springer Berlin / Heidelberg (2005)
15. Tarakanov, A.O., Kvachev, S.V., Sukhorukov, A.V.: A formal immune network and its implementation for on-line intrusion detection. In: Computer Network Security, *Lecture Notes in Computer Science*, vol. 3685, pp. 394–405. Springer Berlin / Heidelberg (2005)
16. Tarakanov, A.O., Skormin, V.A., Sokolova, S.P.: Immunocomputing, Principles and Applications. Springer New York (2003)
17. Tempesti, G.: A self repairing multiplexer-based fpga inspired by biological processes. Ph.D. thesis, École Polythechnique Fédérale de Lausanne (1998)
18. Tyrrell, A.: Computer know thy self!: A biological way to look at fault tolerance. In: 25th Euromicro Conference (EUROMICRO '99), vol. 2, pp. 21–29. 2nd Euromicro/IEEE Workshop on Dependable Computing Systems (1999)
19. Wang, Z., Chakrabarty, K.: Built-in self-test and defect tolerance in molecular electronics-based nanofabrics. In: Journal of Electronic Testing: Theory and Applications, vol. 23, pp. 145–161. Kluwer Academic Publishers (2007)
20. Wunderlich, H.J.: Test and testable design. In: Architecture design and validation methods, pp. 141–190. Springer-Verlag New York (2000)
21. Xilinx Inc.: Xilinx TMRTool, The First Triple Module Redundancy Development Tool for Reconfigurable FPGAs. URL www.xilinx.com

An Organic Computing Approach to Sustained Real-time Monitoring

Rainer Buchty, David Kramer, Wolfgang Karl

Abstract System monitoring is not only the key to system and application optimization, but also provides the fundamental functionality for adaptive and self-configuring systems. In this paper we describe a novel approach to monitoring based on biologically inspired methods, which not only suits traditional requirements in generating a detailed and pristine system state image, but also complies with the dedicated needs of self-configuring systems. As a beneficial side-effect, the proposed monitoring approach is inherently fault-tolerant and scalable with system size.

Keywords: Monitoring, Bio-inspired, Hormone System, Adaptive Computing, Self-X.

1 Introduction

Traditional use of monitoring techniques range from basic system introspection required for sanity checks or load balancing to detailed system traces used for application and architecture optimization. Especially the latter requires detailed and pristine recording of the system state to enable correlation of program code and monitored effects, such as e.g. cache hits and misses.

For these topics, a plethora of techniques and tools has been developed such as hardware counter registers triggering to individual events, profilers, or simulation infrastructures. Each of these methods, however, comes with certain drawbacks. Simulation is only as precise as the underlying simulation model, and typically several magnitudes slower than execution on a real system. It does, however, provide

Universität Karlsruhe (TH)
Institut für Technische Informatik
76128 Karlsruhe, Germany
{buchty|kramer|karl}@ira.uka.de

Please use the following format when citing this chapter:

Buchty, R., Kramer, D. and Karl, W., 2008, in IFIP International Federation for Information Processing, Volume 268; *Biologically-Inspired Collaborative Computing*; Mike Hinchey, Anastasia Pagnoni, Franz J. Rammig, Hartmut Schmeck; (Boston: Springer), pp. 151–162.

the possibility of closest possible introspection and is able to deliver a pristine system view, unaltered by monitoring side-effects.

Simulation is especially hampered by the problem of data size: even short program runs of few milliseconds are able to generate several megabytes of monitoring (trace) data at single-step resolution. Even at a far more coarse-grained monitoring resolution, a program run will easily generate traces in the gigabyte range.

All approaches typically share the way how monitoring data is collected and transported: collection is usually done using fixed, assigned counter registers which are subsequently polled by a monitoring framework. Hence, in real systems, side effects caused by monitoring occur such as interrupt triggering or data transport. When analyzing trace data, these side effects have to be carefully taken into account.

1.1 Adaptive and Self-configuring Systems

Conventional optimization relies on trace generation, off-line analysis of the generated trace data, and manual or semi-automated optimization. For trace generation, monitoring is solely based on pre-defined rules, i.e. the programmer defines upfront which events need to be generated or traced.

This is in contrast to adaptive, self-configuring systems: here, only a start configuration can be provided and it is unknown whether this will suit future system configurations or generate a sufficient amount of monitoring data to enable proper reaction to events. Furthermore, monitoring data must not be collected for off-line introspection, but rather be analyzed and evaluated on-line, adhering to application- and system-defined real-time constraints.

Hence, adaptive systems consist of a closed control loop as illustrated by Figure 1: using monitoring components, the system's current state is derived and evaluated against a destination state defined by so-called objective functions. Using these functions and the adaptive capabilities of the system provided by adaptive components, a refined system configuration is derived and a reconfiguration performed here-on.

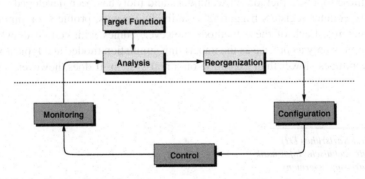

Fig. 1 Adaptive System: Control Loop

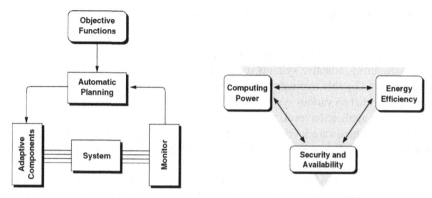

Fig. 2 Adaptive System: Architecture   **Fig. 3** Adaptive System: Objective Functions

Objective functions, however, are usually contradictory. Power vs. performance may serve as an example: almost all dynamically applicable methods to lower power consumption will also decrease performance – and vice versa. Hence, data must be collected and evaluated in real-time to ensure that application demands, as defined through the provided objective functions, are met.

As said before, this entire process is unlike conventional, narrowly focused optimization of e.g. data locality, where only limited information needs to be gathered and correlated, and where correlation and optimization are typically done offline. Instead, an approach not hampered by a fixed corset of pre-defined events and rules is required which instead is tailored towards the needs of adaptive and self-configuring systems.

1.2 Outline

In this paper, we therefore propose a novel monitoring system approach. This approach employs a uniform, flexible method of associative counter registers, subsequently enabling real-time analysis of monitoring data, and minimizing side-effects caused by monitoring data transfer. The approach does not require up-front definition of monitoring rules; instead, a bio-inspired method is used, enabling fault-tolerant event generation, distribution, and evaluation. Therefore, within the scope of an adaptive system, also monitoring itself becomes adaptive and – as a beneficial side-effect of the used method, – easily scalable with system size.

This paper is organized as follows: we will first give an introduction to the specific set-up and needs of adaptive, self-configuring systems illustrated by a novel-bio-inspired architecture. We then discuss existing monitoring approaches and their suitability with respect to real-time capabilities, system influence, and use within adaptive, self-configuring systems, followed by detailed presentation of our monitoring approach, its prototype implementation and results. The paper is closed with the conclusion and outlook.

2 Related Work

Trace generation for optimization purposes and feedback systems as required for self-configuring, adaptive systems require techniques to gather and process system parameters to be able to create a certain sense of self-awareness. These parameters can be collected on various system levels such as lowest hardware level, driver level, OS level, or application level.

On lowest hardware level, performance counters offer some rudimentary monitoring support. They are typically used to profile an application and investigate possible application optimizations, such as enhanced data layout in memory to improve cache use. Modern processor architectures offer so-called event or performance counter registers [2, 6, 11, 12, 18, 10, 17]. Number and use of these registers are dependent on the individual architecture: counter registers are either bound to certain events or can be more or less freely assigned [17, 11]. The majority of existing analysis tools is based on these counter registers.

Counter-based methods typically suffer from four basic limitations [21] which are number of registers, sampling delay, and lack of address profiling. Furthermore, it is not possible to differentiate between events being triggered by speculative and non-speculative execution. False counts from speculation are addressed with the precise event-based sampling (PEBS) of Intel's Pentium 4 architecture [12, 20]. The drawback of this method, however, is increased chip size and influencing the normal system behavior resulting from concurrent regular memory accesses of the currently running program. Similar, but less complex methods are implemented in the IBM's Power architecture [18, 10].

It is a general problem of sampling methods described above, that only system snap-shots are created. Thus, these methods serve only for creation of aggregated statistics. It is usually not possible to selectively pre-compute monitored data already on the monitoring tier, leading to impact resulting from reading and post-processing the counter registers.

While counter registers are well-suited for processor audit, they are not sufficient for flexible, system-wide, and generic monitoring concepts as required for autonomic computing systems. They are fixed and furthermore lack the possibility of pre-processing.

For architectures without such hardware support, monitoring can be achieved using plain software based on profilers inserting function prologues to collect statistical information such as gprof [7]. It is also possible to embed monitoring routines on driver level as demonstrated by the Myrinet-based Shrimp Cluster [13]. A combined hardware/software method was used on the SMiLE monitor for SCI networks [8].

To configure the monitoring system and extract and process monitoring data, higher-level monitoring APIs may be used such as [19], [3], or [14, 15]. Task of these APIs is to decouple monitoring devices from post-processing software by offering an abstract programming interface rather than directly accessing the monitoring hardware.

From the above examples several conclusions can be drawn: *Existing infrastructures in hardware are fixed.* While suitable for their designed task, monitoring infrastructures are too limited for generic use within a self-optimizing system. It is not possible to replace or adapt existing resources. We address this topic by using associative counter arrays instead of fixed event counters as outlined in Section 4.1.

Monitoring systems are application-specific. Powerful monitoring tools exist for various applications, attaching monitoring techniques to various system layers. What is missing, though, is generic support for plugging dedicated monitoring devices into the running system as required. Ideally, monitoring modules can be applied to all system layers through a defined and standardized interface. This is addressed by a uniform event specification as outlined in Section 4.2, enabling unique event identification where this identification also includes where and how an event was generated.

No standardized API exists. So far, several approaches for monitoring APIs exist, However, these are typically bound to certain applications. No uniform, application-independent, and standardized monitoring API being able to report existing monitoring resources, permitting to access these resources, and – with respect to self-organizing systems – enabling reconfiguration exists. We account for this by using bio-inspired mechanisms for event evaluation also discussed in Section 4.2.

3 DodOrg: A Self-configuring Bio-inspired Architecture

For upcoming dynamically changing and self-adjusting systems the conventional method of monitor-data processing is not suitable anymore. Such systems internally feature not one single observer/controller architecture, instead a multitude of individual control loops may coexist creating a hierarchy of de-central, decoupled control loops.

We want to illustrate this with the Digital on-demand Computing Organism for Real-time Systems (DodOrg) [4], a novel computer architecture inspired by biology.

Like biological organisms, DodOrg is hierarchically structured: its hardware is provided by so-called organic processing cells (OPCs). The OPCs provide different capabilities ranging from general-purpose processing to DSP and reconfigurable FPGA blocks, memories, and dedicated I/O cells. All cells are arranged in a grid featuring peer-to-peer connection of the cells.

Mapping an application's task to these cells follows the idea of organ formation in biological organisms: individual OPCs are grouped into work clusters, or organs, using a de-central hormone-inspired middleware. This grouping is based on an individual cell's suitability for application tasks, i.e. its architecture (CPU vs. DSP vs. FPGA), computing power, already assigned work load, and energy demands such as provided vs. required energy.

This is akin to biological organisms where the concentration level of hormones is measured. If a certain threshold is reached, some action is triggered such as e.g. rising blood pressure and heart beat rate. It is also possible, that based on the threshold levels of one or more individual hormones a so-called second messenger, i.e. a new, derived event type, is generated.

Any organic architecture therefore requires a comprehensive, flexible, and adaptive monitoring approach, hence system monitoring is the important issue with self-organizing systems. In particular, the entire system must be constantly evaluated, therefore monitoring within DodOrg spans all system levels, i.e. individual monitors are distributed across the system and are connected to different components and layers.

4 The Monitoring Approach

As can be seen from the previous discussion, monitoring faces several challenges: upcoming systems not only demand a high degree of flexibility with respect to event detection and accumulation, but also require real-time correlation and interpretation of such data. Since the system itself might constantly change, no infrastructure relying on pre-defined events and monitor rules is able to provide necessary adaptivity.

We therefore propose two basic mechanisms for next-generation monitoring infrastructures suitable for both, scalable systems and dynamically, self-adapting systems: the limitation of monitoring resources we address by using so-called associative counters triggering to arbitrary events instead of being hard-wired to a small pre-selection of individual events. This obviously requires self-identifying event coding.

In Section 4.1, we therefore show our proposed encoding scheme, the associative counter design, and beneficial side effects resulting from this approach.

Subsequently, in Section 4.2, we show how this unique event coding matches the biological model of hormones or messengers and how derived mechanisms overcome the necessity for pre-defined monitoring rules, resulting in an inherently flexible and adaptive nature of monitoring data evaluation.

4.1 Associative Counter Arrays

So-called performance counters enable real-time monitoring of event data without requiring to trigger an external monitoring instance with every event occurrence. Instead, a certain amount of events is counted and later an accumulated number is forwarded to a memory buffer where it then can be processed by a higher monitoring instance, e.g. for correlation or visualization.

In contrast to conventional systems, where both, counter size and event association are typically fixed, our monitoring approach features the use of associative counter arrays. Hence, no hard-wired connection between one or more predefined events and a single counter exists; instead, the counters are self-triggering to any event. Introducing a programmable modulo – potentially useful for more infrequent or rare events – enables control of the counter overflow, aiding histogram generation for higher-level monitoring.

Figure 4 illustrates the general construction and work principle: upon event occurrence, the event is caught by the counter array and assigned to the first spare, i.e. unassigned, counter. Subsequent occurrences of this very event will trigger the assigned counter. If no spare counter is available, an assigned counter will be reassigned: in this case, the existing counter value and the event identification are evicted from the counter array, and the referring counter is initialized and assigned to the new event.

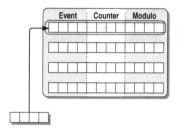

Fig. 4 Associative Counter Array: Principle of Operation

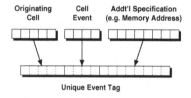

Fig. 5 Unique Event ID Construction

From this explanation it becomes obvious that the associative counter array is a cache memory by nature where the event ID becomes the cache tag and the the event-associated counter resembles the cached data. Hence, likewise replacement strategies including, but not limited to, FIFO, LRU, and LFU approaches.

To make this concept work, a unique and self-defining encoding of system events as shown in Figure 5 is required. This encoding consists of two parts, a local part denoting the event itself (i.e. a memory access) and associated data (the address and read/write flag), and an additional part containing the source of this event, i.e. which CPU performed this operation.

In the following, we will show that such a unique event tag does not only ease event monitoring using associative counter arrays, but also correlation and evaluation of such event monitor data.

4.2 Biologically Inspired Event Communication and Evaluation: The Hormone Concept

Associative counter arrays already provide sufficient flexibility for event detection; in addition to that, also a flexible method of event correlation is required, including the possibility of focus adjustment, i.e. changing the event granularity or "monitoring resolution".

A suitable technique can, again, be found in biological organisms: here, certain events are communicated using so-called messengers, or hormones. The amount of messengers – i.e. events – generated is solely defined by its respective producers, i.e. no central instance commands generation of events, neither does a central instance command event encoding.

In biological organisms, the concentration level of hormones is measured. If a certain threshold is reached, some action is triggered, such as e.g. rising blood pressure and heart beat rate. It is also possible that based on the threshold levels of one or more individual hormones a so-called second messenger, i.e. a new, derived event type, is generated. Messengers are inherently self-defining by their chemical structure; hence, using an encoding mechanism and a method of accumulation as described in Section 4.1, both, hormone concept in general, and concentration triggering in particular can be directly applied to digital systems as illustrated by Figure 6.

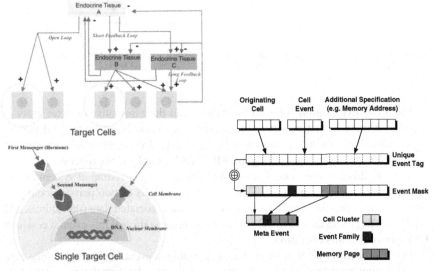

Fig. 6 Hormone Reception, Processing, and Second Messenger Principle

Fig. 7 Mask-based Event ID Evaluation

In such a system, whether to react or not to one or more hormones is solely decided by the receiver. With the absence of central instances therefore the the overall system becomes very flexible and scalable. Further amount of flexibility is introduced by not using event IDs directly, but to apply a simple identification mask and therefore be able to narrow and widen the focus of event accumulation: with this mask, mandatory and "don't care"-fields of the event ID are specified. This mask is then applied to incoming event IDs (using a simple Boolean function). In case of a match, either a counter may be triggered or, mimicking the so-called second-messenger principle, a new event may be sent into the system. Figure 7 shows such an evaluation process.

Since only the interpretation of already generated monitoring data is changed, different monitoring instances might coexist, each interpreting the present monitoring data differently.

The drawback of such an approach is a potentially high communication load imposed by transportation of monitoring data. However, [1] showed that for typical setups the amount of monitoring data can be easily transported in modern communication infrastructures, including networks on chips (NoCs).

5 Prototype Implementation and Results

To fully elaborate the concept, we developed two prototypes: a software prototype is used to demonstrate the suitability of the overall concept, and serves also as a case study for using the proposed concept to enhance an existing communication infrastructure.

In addition, we developed a hardware prototype aimed at monitoring traffic in contemporary high-performance bus-systems employed in current and future multicore systems for core interconnection.

Our **software prototype** targets a sub-problem of distributed and heterogeneous architectures such as DodOrg: in such architectures, memory allocation, access, and access right management becomes crucial. Hence, we developed the concept of so-called Self-aware Memory [5], providing a more intelligent, scalable, and de-central memory management suitable for highly heterogeneous parallel systems with special focus on dynamic, adaptive systems.

Communication within SaM is based on a lightweight, hormone-inspired protocol, where each memory access (allocation request, grant, and read/write access) is therefore encoded as a unique event.

The monitoring approach proposed in this paper was applied to an existing SaM simulation environment [16] where it will be used as the information-gathering entity required for monitoring and correlating memory accesses to steer autonomous defragmentation, locality optimization, and brokering. This environment basically consists of a number of CPUs and memory banks connected through a shared interconnection network.

As a first step towards full self-management, we therefore extended the existing simulation infrastructure to provide associative counter arrays in each component to be able to detect incoming and outgoing accesses for each component, hence, monitoring individual commands to record typical request/response behavior taking place within the SaM protocol.

The setup was verified using a basic functionality test. For this test, we monitored randomly generated memory accesses, proving the functionality of our monitoring approach. This monitor was then subsequently used for further development of the

SaM memory allocation protocol, where we successfully verified and measured the behavior of the protocol additions and alterations.

With this setup, we were easily able to introduce and verify a new, more efficient allocation mechanism replacing the formerly used strategy by simply triggering to the individual access messengers (i.e. event IDs of individual accesses); while certainly overkill for just protocol optimization, the software prototype is, however, the initial and vital step to explore self-management, leading to fully autonomous self-optimization as required within the SaM context.

We also developed a **hardware prototype** to prove that our approach is suitable for high-performance communication technologies required for on-chip multicore connectivity as e.g. employed within DodOrg, and give a first estimation of the associated hardware costs. We therefore chose HyperTransport [9] as a state-of-the-art interconnection system, using an existing interface core [22], which was extended by an associative counter array to monitor memory accesses from CPU to memory. The array consists of 64 individual 8-bit counters with a tag size of 40 bits.

Targeting a Xilinx Virtex4FX100, this monitor accounts for about 4% of logic use (slices), mostly holding the access logic, and 6% of on-chip memory storage (RAMB16) for the associative counter array; compared to the original, unaltered core the monitor-equipped core shows an increase of 36% in logic and 33% in memory.

6 Conclusion

In this paper we presented a novel approach to sustained real-time monitoring. The approach not only introduces increased flexibility, but also addresses the specific topic of dynamically changing and self-adapting systems.

The approach employs associative counter arrays, introducing the flexibility required for such systems as counters are no longer bound a single event or predefined small group of events. Doing so requires the introduction of unique event IDs so that events are inherently self-defining. Applying a simple match mask enables scaling of the monitoring resolution so that counters may react to individual events or configurable groups of events, such as e.g. events coming from a distinct source or memory accesses with configurable address granularity.

This concept is biologically inspired and based on the hormone concept where solely the designated receiver decides whether to receive and how to react to hormones. No centralized monitoring instance exists, instead a distributed systems of producers (cells emitting hormones) and receivers (cells receiving hormones) exist. Hence, the entire system is inherently scalable and fail-safe.

To evaluate the concept, we extended a simulation infrastructure where we applied this style of monitoring. It required very little programming work and proved very successful in the process of expanding, optimizing, and verifying a communication protocol for de-central memory management and access.

We furthermore developed a hardware prototype to demonstrate the general suitability for high-performance real-time systems, and evaluated the hardware requirements such as chip area and ease of integration into existing hardware infrastructures.

By the outcome of this work we are convinced that the proposed method does provide the necessary flexibility and ease of use as required for dynamical and adaptive systems. We were able to show the required real-time capabilities, and the hardware requirements proved to be modest so that they can be easily integrated in existing interface structures such as the used HTX core.

We will therefore further pursue this work and not only refine the unique event identifier scheme and optimize our existing prototypes, but also be soon able to test our concept against real-world examples, running dedicated benchmark suites. Doing so will generate more universal results with respect to hardware requirements and communication overhead.

7 Acknowledgements

The software and hardware prototypes were implemented by Luis Mariano Guerra (software) and Bastian Molkenthin (hardware) as part of their semester projects.

References

1. Uwe Brinkschulte Alexander von Renteln. Reliablity of an Artificial Hormone System with Self-X Properties. In *Parallel and Distributed Computing and Systems*, Cambridge, Massachusetts, USA, November 19-21 2007.
2. AMD. AMD Athlon processor, x86 Code Optimization Guide. 2002.
3. J.M. Anderson, L.M. Berc, J. Dean, S. Ghemawat, M.R. Henzinger, S.-T.A. Leung, R.L. Sites, M.T. Vandevoorde, C.A. Waldspurger, and W.E. Weihl. Continuous profiling: Where have all the cycles gone? In *Proceedings of the 16th ACM Symposium on Operating Systems Principles*, Oct 1997.
4. Jürgen Becker, Kurt Brändle, Uwe Brinkschulte, Jörg Henkel, Wolfgang Karl, Thorsten Köster, Michael Wenz, and Heinz Wörn. Digital On-Demand Computing Organism for Real-Time Systems. In Wolfgang Karl, Jürgen Becker, Karl-Erwin Großpietsch, Christian Hochberger, and Erik Maehle, editors, *Workshop Proceedings of the 19th International Conference on Architecture of Computing Systems (ARCS'06)*, volume P81 of *GI-Edition Lcture Notes in Informatics (LNI)*, pages 230–245, March 2006.
5. Rainer Buchty, Oliver Mattes, and Wolfgang Karl. Self-aware Memory: Managing Distributed Memory in an Autonomous Multi-Master Environment. In *The 2008 International Conference on Architecture of Computing Systems (ARCS 2008)*, Dresden, Germany, February 25-28 2008.
6. Compaq Computer. Alpha 21264 Microprocessor Hardware Reference Manual.
7. J. Fenlason and R. Stallman. GNU gprof: The GNU Profiler. 1997.
8. Robert Hockauf, Wolfgang Karl, Markus Leberecht, Michael Oberhuber, and Michael Wagner. Exploiting Spatial and Temporal Locality of Accesses: A New Hardware-Based Monitoring Approach for DSM Systems. In David Pritchard and Jeff Reeve, editors, *Euro-Par'98 Parallel*

Processing, 4th International Euro-Par Conference, Southampton, UK, September 1-4, 1998 Proceedings, volume 1470 of Lecture Notes in Computer Science, Berlin, September 1998. Springer Verlag.

9. HyperTransport Consortium. HyperTransport: Low latency Chip-to-Chip and beyond Interconnect. http://www.hypertransport.org.

10. IBM. PowerPC 740/PowerPC 750 RISC Microprocessor User's Manual. 1999.

11. Intel. Intel Itanium Architecture Software Developer's Manual. 2000.

12. Intel. Intel Architecture Software Developer's Manual Volume 3: System programming Guide. 2002.

13. C. Liao, M. Martonosi, and D.W. Clark. Performance monitoring in a myrinet-connected shrimp cluster. In Proceedings of the International Conference on Measurement and Modeling of Computer Systems (Sigmetrics'98), Aug 1998.

14. T. Ludwig, R. Wismüller, V. Sunderam, and A. Bode. OMIS – On-line Monitoring Interface Specification (Version 2.0). Shaker Verlag, Aachen, Germany, 1997. ISBN 3-8265-3035-7.

15. Thomas Ludwig and Roland Wismüller. OMIS 2.0 — A Universal Interface for Monitoring Systems. In M. Bubak, J. Dongarra, and J. Wasniewski, editors, Recent Advances in Parallel Virtual Machine and Message Passing Interface, volume 1332 of Lecture Notes in Computer Science, pages 267–276, November 1997.

16. Oliver Mattes. Developing a decentral memory management for distributed systems (self-aware memory). Master's thesis, Universität Karlsruhe (TH), 2007.

17. Sun Microsystems. Ultra-SPARC IIi User's Manual. 1997.

18. Motorola. MPC7450 RISC Microprocessor Familiy User's Manual. 2001.

19. P.J. Mucci, S. Browne, C. Deane, and G. Ho. PAPI: A portable interface to hardware performance counters. In Proceedings of the Department of Depense HPCMP User Group Conference, Jun 1999.

20. B. Sprunt. Pentium 4 performance-monitoring features. In IEEE Micro, pages 72–82, Jul/Aug 2002.

21. B. Sprunt. The basics of performance-monitoring hardware. In IEEE Micro, pages 64–71, Jul/Aug 2002.

22. Ulrich Brüning et al. HTX Board Universal Reference Design. http://www.hypertransport.org/products/productdetail.cfm?RecordID=75.

A Case Study in Model-driven Synthetic Biology

David Gilbert, Monika Heiner, Susan Rosser, Rachael Fulton, Xu Gu and Maciej Trybiło

Abstract We report on a case study in synthetic biology, demonstrating the model-driven design of a self-powering electrochemical biosensor. An essential result of the design process is a general template of a biosensor, which can be instantiated to be adapted to specific pollutants. This template represents a gene expression network extended by metabolic activity. We illustrate the model-based analysis of this template using qualitative, stochastic and continuous Petri nets and related analysis techniques, contributing to a reliable and robust design.

1 Motivation

One of the greatest challenges in modern bioscience is arguably the development of techniques for the engineering of living systems in a rigorous manner. This is the domain of the emerging discipline of "Synthetic Biology" [HP06], which can be defined as the design and construction of new biological parts, devices, and systems, as well as the re-design of existing natural biological systems for useful purposes [Syn08]. One aspect of Synthetic Biology which distinguishes it from conventional genetic engineering is a heavy emphasis on the development of foundational technologies that make the engineering of biology easier and more reliable.

David Gilbert, Rachael Fulton, Xu Gu, Maciej Trybiło
Bioinformatics Research Centre, University of Glasgow, Glasgow G12 8QQ, Scotland, UK, e-mail: drg@brc.dcs.gla.ac.uk, 0307842f@student.gla.ac.uk, gux@brc.dcs.gla.ac.uk, trybilom@dcs.gla.ac.uk

Monika Heiner
Department of Computer Science, Brandenburg University of Technology, Postbox 10 13 44, 03013 Cottbus, Germany, e-mail: monika.heiner@tu-cottbus.de

Susan Rosser
Institute of Biomedical and Life Sciences, University of Glasgow, Glasgow G12 8QQ, UK, e-mail: s.rosser@bio.gla.ac.uk

Please use the following format when citing this chapter:

Gilbert, D., Heiner, M., Rosser, S., Fulton, R., Gu, X. and Trybilo, M., 2008, in IFIP International Federation for Information Processing, Volume 268; *Biologically-Inspired Collaborative Computing*; Mike Hinchey, Anastasia Pagnoni, Franz J. Rammig, Hartmut Schmeck; (Boston: Springer), pp. 163–175.

We report on a case study in synthetic biology [Gla07], demonstrating the model-driven construction of a completely novel type of self-powering electrochemical biosensor, called ElectrEcoBlu. The novelty lies in the fact that the response signal is an electrochemical mediator which enables electrical current to be generated in a microbial fuel cell. ElectrEcoBlu functions as a biosensor for a range of important and widespread environmental organic pollutants which stimulate the biosensor to produce its own electrical power output. The system has the potential to be used for self-powered long term in situ and online monitoring with an electrical readout.

Our approach exploits a range of state-of-the art modelling techniques [GHL07] to guide the design and construction of this novel synthetic biological system in order to ensure that its behaviour is reliable and robust under a variety of conditions. This was facilitated by the entire team - molecular biologists and engineers/modellers - working in an integrated laboratory environment, using Petri nets as a communication means and following an iterative construction process as given in Fig. 1. An essential result of the design process is a general template of a biosensor, which can be instantiated to be adapted to special pollutants. This template represents a gene expression network extended by metabolic activity. We demonstrate the model-based analysis of this template, and by this way of the design of the biosensor, using qualitative, stochastic and continuous Petri nets and related analysis techniques.

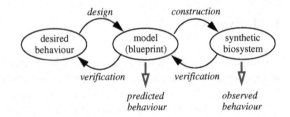

Fig. 1 Model-driven synthetic biology. Computer modelling and analysis guides the design and construction in order to ensure behaviour, which is reliable and robust under a variety of conditions.

2 Biochemical Context

Public concern and legislation are demanding better environmental control and monitoring of pollutants. Biosensors are being developed in the fields of environment, bioprocess control, food, agriculture, military, and medical industries. Biosensor sensitivity and selectivity depend essentially on the properties of the biorecognition elements to be used for analyte binding.

The discovery of transcriptional activators and their corresponding promoter sequences has made possible the development of bacterial biosensors for pollutants. Modified cell biosensors are constructed by fusing a reporter gene (an enzyme or a fluorescent protein e.g. GFP) to a promoter element that is induced by the presence of a target compound. In the presence of an organic contaminant the transcriptional activator changes its three dimensional structure, becoming operative, and transcription of the reporter gene is enhanced. The gene is transcribed to form mRNA which is then translated into a protein which performs the biochemical activity (in the case of enzymes) or fluoresces (e.g. GFP). The resulting increase in reporter gene product is then detected by measuring the activity of the reporter enzyme or the fluorescence of the reporter protein. Thus, under appropriate conditions, a direct correlation between contaminant concentration and reporter product can be established.

One holy grail of environmental biosensors is to create a system which can be left in the field continuously monitoring and remotely sending electronic signals back to a computer. One major problem is how to power such a device so that frequent expensive battery changes are not necessary. One possible source of renewable energy for powering biosensor devices are microbial fuel cells (MFC) in which microorganisms oxidize compounds such as glucose, acetate or wastewater. The electrons gained from this oxidation are transferred to an electrode. In the past, external, expensive, soluble redox mediators have consistently been added to MFCs to enhance electron transfer. *Pseudomonas aeruginosa* has been shown to produce its own electron transporters, pyocyanin (PYO), which can function as electron-carrying redox mediators increasing electrical power output of MFCs.

Our project aimed to use a synthetic biology approach to combine the production of an environmental biosensor for economically important industrial environmental pollutants with a microbial fuel cell which can produce its own electricity. The intention is that the cells will recognise the presence of a pollutant via a modular interchangeable range of pollutant-specific transcriptional activator proteins and enhance electricity generation in a microbial fuel cell by inducing genes for the synthesis of the electron mediator PYO which function as novel reporter genes.

The *recognition element* of the designed biosensor system is a pollutant responsive transcriptional activator XylR (DntR) which binds the important environmental pollutant toluene (salicylate). The *reporter element* of the biosensor consists of the enzymes S-adenosylmethionine-dependent N-methyltransferase (PhzM) and flavin-dependent hydroxylase (PhzS) which convert the precursor compound phenazine-1-carboxamide (PCA) to PYO in the biosynthetic pathway cloned from *Pseudomonas aeruginosa* into *E. coli* and a non-pathogenic Pseudomonad strain.

The molecular biologists of our team constructed an initial diagram to describe the system, using a fairly informal graphical syntax, see Figure 2. The generic form of the transcription factor ('tf' for the gene, and 'TF' for the protein product) represents both XylR (toluene detecting) and DntR (salicylate detecting). In outline, essential steps that we used to develop and refine our model are:

1. Simplification by abstracting away the mRNA, thus combining transcription and translation.
2. Summarizing pollutant-specific transcriptional activator proteins under the term TF.
3. Combining the PhzM and PhzS components to give one step from PCA to PYO.
4. Developing a variant of the model with a positive feedback loop (pfb).

By doing so, we obtain a gene expression network, extended by metabolic activity, i.e. the model combines different abstraction levels: gene activity as well as metabolic activity, in the style of [vHNM⁺00]. This represents deliberately a minimal model concentrating on the most essential facts necessary to investigate the system's signal/response behaviour. To be able to analyse the system before having constructed it, we are going to apply formal modelling techniques allowing the computer-based evaluation of the system under construction.

Fig. 2 The general biosensor scheme in two versions: without/with the positive feedback (pfb). Thick arrows represent protein coding genes, thin right-angle arrows represent promotors optionally labelled with the transcription factor, and thin straight arrows represent biochemical reactions. Note that the first instance of the 'tf' protein coding gene is constitutively expressed.

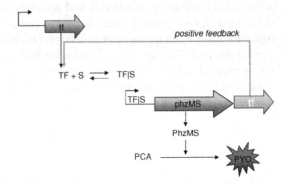

3 Framework

We have used a framework [GHL07] which integrates the qualitative, stochastic and continuous paradigms, as a basis for our overall approach to modelling and analysing the biosynthetic pathways, compare Fig. 3. Each perspective adds its contribution to the understanding of the system, thus the three approaches do not compete, but complement each other.

In summary, the qualitative time-free description is the most basic one, with discrete values representing numbers of molecules or levels of concentrations. The qualitative description abstracts over two timed, quantitative models. In the stochastic description, discrete values for the amounts of species are retained, but a stochastic rate is associated with each reaction. The continuous description models amounts of species using continuous values and associates a deterministic rate with each reaction, which now occurs continuously. These two time-dependent models can be mutually approximated by propensity (hazard) functions belonging to the stochastic world; see [GHL07] for more details.

This framework can be applied to a variety of formalisms; we specify stochastic models by stochastic Petri nets defining reaction rate equations (RREs), and continuous models by continuous Petri nets defining ordinary differential equations (ODEs). In the following we assume basic knowledge in the standard Petri net terminology; see e.g. [BK02, DA05, MBC⁺95] for introduction and related definitions.

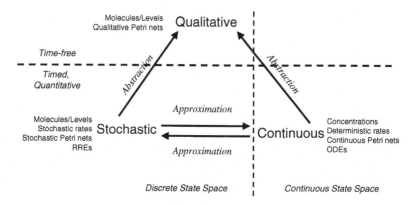

Fig. 3 Conceptual framework of our computational methods.

4 Qualitative Approach

Modelling From the graphical representation of the system given in Figure 2 and its accompanying explanations we derive a qualitative Petri net describing a general kernel biosensor template, and also one possible extension with pfb, see Figure 4. This template and the pfb variant may be instantiated according to Table 1, creating dedicated biosensors for different pollutants. Other variants are possible, including switches and clamps, which we do not discuss here for lack of space.

The Petri net represents an extended gene expression network comprising transitions of various abstraction levels:

- gene expression: TF expression, reporter expression;
- association/deassociation: TFS association, TFS deassociation;
- enzymatic reaction: response production;
- degradation: TF/TFS/reporter/response degradation.

The transition *TF expression* is an input transition (transition without preplaces), modelling a constitutively expressed transcription factor, i.e. a gene which is constantly active. The degradation transitions are output transitions (transitions without postplaces). They model the fact that species naturally degrade, i.e. their concentration diminishes if they are not produced continuously.

The two essential components of a biosensor are easily identified: the recognition element (upper part), and the reporter element (lower part), both coupled by the TFS complex. In the recognition element, the signal (pollutant) forms a complex (TFS) with a constitutively expressed transcription factor (TF). The TFS complex may accelerate the TF expression, thus facilitating faster TFS association; in this case we get a pfb. In the reporter element, the two read arcs (having a black dot as arrow head) reflect the signalling cascade: TFS → reporter → response. In the following we analyse the template, and by this way all its instances.

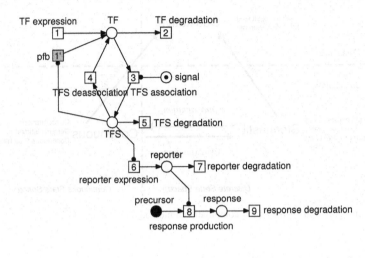

ORD	PUR	HOM	NBM	CSV	SCF	CON	SC	FTO	TFO	FPO	PFO	NC
Y	N	Y	Y	N	N	Y	N	Y	Y	N	N	ES
DTP	CPI	CTI	SCTI	SB	k-b	1-b	DCF	DSt	DTr	LIV	REV	
Y	N	Y	N	N	N	N	-	N	N	Y	-	

Fig. 4 Qualitative Petri net(s) of the biosensor template. The kernel system is given with white transitions, and the positive feedback (pfb) variant has an additional transition indicated in grey. The place given in black models the precursor, which is assumed to be available in sufficiently large amounts; hence it is neglected in the model analyses. The two-lines result vector given at the bottom summarizes the main qualitative analysis results; see [HGGH07] for more details.

Table 1 Possible instantiations for the places in the biosensor template.

place	instances
signal	*toluene, salicylate*
TF	*XylR, DntR*
TFS	*XylR|S, DntR|S*
reporter	*PhzMS* as combination of *PhzM* and *PhzS*
precursor	phenazine-1-carboxamide (*PCA*)
response	pyocyanin (*PYO*)

Analysis Having established initial confidence in the model behaviour by playing the token game, both systems were formally analysed. We list here the most essential analysis results only. For a summary see the two-lines result vector given at the bottom of Figure 4, for more explanatory details see [HGGH07].

The Petri net has input transitions and output transitions, i.e. it is an open system. Input transitions are always enabled, therefore they are able to fire arbitrarily often, making the Petri net unbounded. Consequently, the Petri net is not covered by P-invariants (CPI). Actually, there is only one minimal P-invariant, which comprises merely the place *signal*. That means that the token number on this place never changes under any firing, reflecting the model assumption that the signal (pollutant) is constantly there at a strength as chosen by the initial marking. Therefore, this place requires at least one token in the initial marking to allow its posttransition to fire. On the contrary, all other places - empty in the initial marking - are unbounded, i.e. the token number may rise to infinity if we consider the model under any timing behaviour. The proof of boundedness under given timing constraints, as e.g. by determining a steady state, is left to the quantitative analyses, see Section 5.

Having the initial marking, we consider liveness. The Petri net is ordinary, i.e. all arc weights are equal to 1, and the net belongs to the structural net class Extended Simple (ES). The Petri net with the given initial marking has the deadlock trap property (DTP). The DTP involves liveness for ordinary ES nets. Because the net is live, there are no dead transitions and no dead states. That basically means that all reactions will take place forever. Because the net is ES, the liveness is guaranteed for any timing constraints.

We compute the T-invariants to get the subprocesses, from which the whole infinite system behaviour is comprised. The Petri net without pfb is covered by the following minimal T-invariants (CTI), all enjoying an obvious biological meaning:

$y_1 = \{TF\ expression, TF\ degradation\}$,

$y_2 = \{TF\ expression, TFS\ association, TFS\ degradation\}$,

$y_3 = \{TFS\ association, TFS\ deassociation\}$,

$y_4 = \{reporter\ expression, reporter\ degradation\}$,

$y_5 = \{response\ production, response\ degradation\}$.

The Petri net with pfb has additionally the following two T-invariants (the counterparts to y_1, y_2, replacing $TF\ expression$ by pfb):

$y_6 = \{pfb, TF\ degradation\}$,

$y_7 = \{pfb, TFS\ association, TFS\ degradation\}$.

One of the benefits of using the qualitative approach at this early stage of system design was that the systems could be modelled and analysed without any quantitative parameters. Moreover the qualitative step helps in identifying suitable initial markings and potential quantitative analysis techniques.

5 Quantitative Approaches

Modelling To transform the validated qualitative Petri net into quantitative ones, we need to assign to all reactions their rate functions, which generally employ the current state of the reactions' substrates, or - in Petri net terms - the current marking of the transitions' preplaces. Table 2 gives for each reaction (transition): the reaction equation, the rate function and the involved rate constant(s). The rate functions are used in the stochastic model as the propensity (hazard) functions, determining the current stochastic firing rates, and in the continuous model as the deterministic rate functions, determining the current deterministic firing rates. The conversion of stochastic and deterministic rate constants into each other is well understood, see e.g. [Wil06], especially it holds that they are equivalent for first-order reactions.

Table 2 The reaction equation, rate function, and rate constant(s) for each reaction (transition [a]). The running numbers correspond to the transition numbers in Figure 4. For better readability we use the abbreviations of the instances employed, compare Table 1, and s for the signal. For the concrete values or value ranges of the rate constants see [HGGH07].

#	reaction equation	rate function [b]	rate constant
1	$\phi \rightarrow TF$	c_1	$c_1 = \alpha_{TF}$
1', pfb	$\phi \rightarrow TF$	$(c_{11} \cdot TFS)/(c_{12} + TFS)$	$c_{11} = \beta_{TF}, c_{12} = \gamma_{TF}$
2	$TF \rightarrow \phi$	$c_2 \cdot TF$	$c_2 = \delta_{TF}$
3	$TF + s \rightarrow TFS$	$c_3 \cdot s \cdot TF$	$c_3 = \beta_{TFS}$
4	$TFS \rightarrow FT$	$c_4 \cdot TFS$	$c_4 = k_d$
5	$TFS \rightarrow \phi$	$c_5 \cdot TFS$	$c_5 = \delta_{TFS}$
6	$\phi \rightarrow PhzMS$	$(c_{61} \cdot TFS)/(c_{62} + TFS)$	$c_{61} = \beta_{PhzMS}, c_{62} = \gamma_{PhzMS}$
7	$PhzMS \rightarrow \phi$	$c_7 \cdot PhzMS$	$c_7 = \delta_{PhzMS}$
8	$\phi \rightarrow PYO$	$c_8 \cdot PhzMS$	$c_8 = \alpha_{PYO}$
9	$PYO \rightarrow \phi$	$c_9 \cdot PYO$	$c_9 = \delta_{PYO}$

[a] The preplaces of a transition correspond to the reaction's substrates, and its postplaces to the reaction's products.
[b] Reactions 1', 6 employ Michaelis-Menten kinetics, while all others follow the mass action kinetics.

Finding the rate constants proved to be a difficult and time consuming process. It involved both searches for scientific papers and also discussions with the biologists on our team. Sometimes the exact value for a parameter could not be found due to lack of published material on the reactions involved, however the biologists managed to identify a suitable range of values between which the parameter would fall. The values for the TF were estimated using average values for bacterial transcription factors, those for PhzMS using standard rates for similar proteins, and those for PYO using rates from the literature [OAM+03, PGS+07].

In the following we illustrate the strength of quantitative approaches by selected examples with special emphasis on sensitivity analysis, which aims at the identification of those parameters to which a system is sensitive; i.e. small changes in a parameter's value significantly affect the system behaviour. We start with the stochastic approach to exclude eccentric system behaviour caused by stochastic noise, before considering the averaged behaviour in the deterministic continuous approach.

Stochastic analysis The class of stochastic Petri nets [BK02, MBC+95] associates an exponentially distributed firing rate (waiting time) with each transition, specified by a firing rate λ. Generally, this state-dependent firing rate is defined by a propensity (hazard) function. Table 2 provides the details which permit reading the net in Figure 4 as a stochastic Petri net, specifying at the same time RREs.

The unboundedness of the underlying qualitative model precludes the use of all standard Markov analysis techniques, which are based on the state transition matrix. Applying Gillespie's exact simulation algorithm [Gil77] produces data describing the dynamic evolution of the biological system over time. Note that the template describes the model for one cell; the organism that we use as the 'chassis' for our synthetic system is *E. Coli*, which is unicellular. However the bacteria exist in colonies comprising many cells, each of which contributes to the total production of the response. Thus we have carried out simulations for different size colonies in order to investigate the effect on the observed behaviour, under the reasonable assumption that there is no interaction between individuals in the colony. Figure 5 shows the output of the response (PYO) over time for a signal $s = 10\mu$M. Each graph represents a different number of cells being simulated (1, 10, 100 and 1000), averaged over 10 runs. The noise decreases as the number of cells increases, thus behaviour of the stochastic model approaches that of the deterministic model (see Figure 7). Moreover it is obvious that the system reaches a steady state in all shown cases, determining the value at which the response saturates.

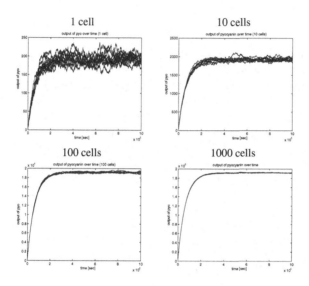

Fig. 5 Diagrams displaying the response (PYO) in number of molecules to a signal $s = 10\mu$M over time produced by simulations for 1, 10, 100 and 1000 cells, averaged over 10 runs.

The goal was to construct a biosensor that would yield a graded signal response. Due to the difficulties experienced in obtaining data, simulations were compared to that in [WWR+98], where a similar system was investigated. In that paper a graded response of the luminescent output was measured over different signal concentrations. The main parameter that affects this was found to be γ_{PhzMS}. Using the plot from [WWR+98], we investigated different values of γ_{PhzMS} ranging from 170 to 500 μM. A graph for response over signal for each of the examined 5 values of γ_{PhzMS} (10 cells) is given in Fig. 6 with standard deviation intervals of response at each value of signal. A lower value of γ_{PhzMS} gives a more graded response, which is consistent with the results reported in [WWR+98].

For more examples and results of the stochastic analyses see [FM07]. We continue with the computationally less expensive continuous approach considering the averaged behaviour.

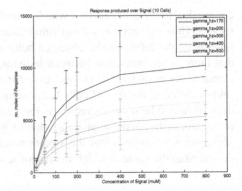

Fig. 6 Diagram showing increasingly graded response to the signal for decreasing values of γ_{PhzMS} (10 cells).

Continuous analysis In a continuous Petri net [DA05] the marking of a place is no longer an integer, but a non-negative real number, and transitions fire continuously according to the deterministic rate functions. Assigning these rate functions to all transitions, see Table 2, and reading the net in Figure 4 as a continuous Petri net, generates the ODEs as given in the equations (1) - (4). The last term in equation (1) corresponds to the pfb transition (given in grey in Figure 4).

$$\dot{TF} = \alpha_{TF} - \delta_{TF} \cdot TF - \beta_{TFS} \cdot s \cdot TF + k_d \cdot TFS + \beta_{TF} \frac{TFS}{\gamma_{TF} + TFS} \quad (1)$$

$$\dot{TFS} = \beta_{TFS} \cdot s \cdot TF - k_d \cdot TFS - \delta_{TFS} \cdot TFS \quad (2)$$

$$\dot{PhzMS} = \beta_{PhzMS} \frac{TFS}{\gamma_{PhzMS} + TFS} - \delta_{PhzMS} \cdot PhzMS \quad (3)$$

$$\dot{PYO} = \alpha_{PYO} \cdot PhzMS - \delta_{PYO} \cdot PYO \quad (4)$$

Simulating the continuous Petri net, i.e. solving numerically the underlying system of ODEs, we get data as given in Figure 7. Here we continue the steady state analysis, comparing the kernel system against its variant in order to determine the influence of the pfb. We see clearly that in both graphs the initial concentrations of PhzMS and PYO both start at zero. Once the reactions have begun (triggered by a pollutant) they increase until they reach a steady state. The interesting point to note is that the production of PYO and PhzMS is much higher in the system with pfb. In quantitative terms, the model with pfb gives around 30% gain. This result allows interesting insights into possible design decisions for the system under construction.

Further, we applied a variant of multi-parametric sensitivity analysis (MPSA) in order to determine those parameters which play a significant role in distinguishing the behaviour between the two models (basic version, and with pfb). Parameter γ_{PhzMS} turned out to be the most sensitive. Thus, in order to refine our comparison we would first try to narrow the value range for this parameter. Such information permits prioritization of costly and time consuming wet-lab experiments.

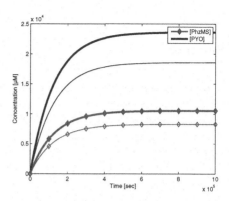

Fig. 7 Dynamic behaviour of the continuous Petri nets allowing comparison of the two system variants. The thin curves belong to the kernel model, and the thick curves to the model with pfb.

6 Tools

The Petri net models were built using the Snoopy [Sno, HRS08] software which is a software tool to design and animate hierarchical graphs, among others Petri nets. It supports qualitative, stochastic and continuous Petri nets, and incorporates the exact Gillespie algorithm for stochastic nets and a variety of ODE solvers for continuous nets. Snoopy provides export to various analysis tools as well as SBML import and export.

The analysis of qualitative Petri nets was performed using Charlie [Cha] which can perform analysis of structural properties, invariant analysis, reachability graph based analysis, and generate visualisations of reachability and coverability graphs. These two tools were developed at the Brandenburg University of Technology, Cottbus.

A specialised Gillespie-style "slow scale stochastic simulation algorithm" [CGP05] was coded in Matlab [MAT] in order to produce the stochastic simulations of the bio-sensor and the graphs in Figure 5 and Figure 6. Fast reactions are computed separately to the slow reactions in order to avoid the slow behaviour of a standard Gillespie-style algorithm. This approach was required due to the properties of the sensor system, where the binding and unbinding reactions of TFS are over six orders of magnitude faster than the other reactions.

The multi-parametric analysis of the model to determine those parameters which play a significant role in distinguishing the behaviour between the two models was performed using the Minicap package [Fri] implemented in Matlab and exploiting its specialised ODE solvers MATLAB [SR97].

The SBML version of the model was generated using the BioNessie [Bio] bio-chemical pathway simulation and analysis tool developed at the University of Glasgow.

7 Summary

The formal modelling and analysis mechanisms of Petri nets have been used in a synthetic biology project to design a completely novel type of self-powering electrochemical biosensor, called ElectrEcoBlu. The novelty lies in the fact that the response signal is an electrochemical mediator, which enables electrical current to be generated in a microbial fuel cell. The work was facilitated by a team of molecular biologists and engineers/modellers working in an integrated laboratory environment, using Petri nets as a communication means.

The 'ElectrEcoBlu' project was carried out as part of the activities of the University of Glasgow's team in the 2007 international Genetically Engineered Machines (iGEM) Synthetic Biology competition, for which they won the Environment and Sensor prize and a gold medal [iGE07].

The outcome so far is the design of a general template of a biosensor, which provably corresponds in various aspects to the desired behaviour. In the next step, the engineered cells will be constructed, i.e. they will be placed in a MFC and the electricity generated under varying conditions and pollutant concentrations will be measured. It is anticipated - supported by the model-based analyses - that the presence of the pollutant toluene would result in an enhanced reporter gene product giving rise to the electron mediator pyocyanin. This in turn is expected to increase the efficiency of electricity production resulting in a measurable electronic signal proportional to the concentration of pollutant.

The model in its three versions and more related material are available at: www.brc.dcs.gla.ac.uk/iGEM/2007.

Acknowledgements We wish to acknowledge David Leader for his valuable contribution to the determination of rate parameters, and Christine Harkness for her initial work on the continuous Petri net model. The entire Glasgow iGEM team contributed to the work of the project, including students Toby Friend, Mai-Britt Jensen, Karolis Kidykas, Martina Marbà, Linsey McLeay, Christine Merrick, Maija Paakkunianen and Scott Ramsay and staff David Forehand, Gary Gray, Margaret Jackson, Raya Khanin, Emma Travis, and Gabriela Kalna. Rachael Fulton was supported by a stipend from the Nuffield foundation. In addition we acknowledge financial support by Scottish Enterprise, the European Union NEST program, the Carnegie Trust, and the University of Glasgow as well as sponsorship from Merk and Anachem.

References

[Bio]   BioNessie. A biochemical pathway simulation and analysis tool. University of Glasgow, www.bionessie.org.

[BK02]   F. Bause and P.S. Kritzinger. *Stochastic Petri Nets*. Vieweg, 2002.

[CGP05]   Y. Cao, D.T. Gillespie, and L.R Petzold. The slow-scale stochastic simulation algo-
           rithm. *Journal of Chemical Physics*, 122(1):014116+, 2005.

[Cha]   Charlie. A Tool for the Analysis of Place/Transition Nets. http://www-
           dssz.informatik.tu-cottbus.de/software/charlie/charlie.html.

[DA05]   R. David and H. Alla. *Discrete, Continuous, and Hybrid Petri Nets*. Springer, 2005.

[FM07]   R. Fulton and M. Marba. A Stochastic Model for a General Biosensor.
           www.brc.dcs.gla.ac.uk/iGEM/2007, Bioinformatics Research Centre, University of
           Glasgow, UK, 2007.

[Fri]   T. Friend. Minicap - Multi Parameter Sensitivity Analysis. Glasgow iGEm Team,
           http://parts.mit.edu/igem07/index.php/Glasgow/Modeling.

[GHL07]   D. Gilbert, M. Heiner, and S. Lehrack. A unifying framework for modelling and
           analysing biochemical pathways using Petri nets. In *Proc. CMSB 2007*, pages 200–
           216. LNCS/LNBI 4695, Springer, 2007.

[Gil77]   D.T. Gillespie. Exact stochastic simulation of coupled chemical reactions. *The Jour-
           nal of Physical Chemistry*, 81(25):2340–2361, 1977.

[Gla07]   Glasgow University Team at iGEM - International Genetically Engineered Machine
           Competition. http://www.brc.dcs.gla.ac.uk/iGEM/2007/, Cited 20 Jan 2008, 2007.

[HGGH07]   C. Harkness, D. Gilbert, X. Gu, and M. Heiner. The use of Petri nets in the
           Glasgow iGEM project: ElectrEcoBlu – a Self-powering Electrochemical Biosen-
           sor. www.brc.dcs.gla.ac.uk/iGEM/2007, Bioinformatics Research Centre, University
           of Glasgow, UK, 2007.

[HP06]   M. Heinemann and S. Panke. Synthetic biology - putting engineering into biology.
           Bioinformatics, 22(22):2790–2799, 2006.

[HRS08]   M. Heiner, R. Richter, and M. Schwarick. Snoopy - A Tool to Design and Ani-
           mate/Simulate Graph-Based Formalisms. In *Proc. PNTAP 2008, associated to SIMU-
           Tools 2008*. ACM digital library, 2008.

[iGE07]   iGEM - International Genetically Engineered Machine Competition, MIT.
           http://parts.mit.edu/igem07/, Cited 20 Jan 2008, 2007.

[MAT]   MATLAB. High-level language and interactive environment. MatWorks,
           www.mathworks.com.

[MBC+95]   M. Ajmone Marsan, G. Balbo, G. Conte, S. Donatelli, and G. Franceschinis. *Mod-
           elling with Generalized Stochastic Petri Nets*. Wiley Series in Parallel Computing,
           John Wiley and Sons, 1995. 2nd Edition.

[OAM+03]   Y.Q. O'Malley, M.Y. Abdalla, M.L. McCormick, K.J. Reszka, G.M. Denning, and
           B.E. Britigan. Subcellular localization of Pseudomonas pyocyanin cytotoxicity in
           human lung epithelial cells. *Am J Physiol Lung Cell Mol Physiol*, 284(2):L420–430,
           2003.

[PGS+07]   J.F. Parsons, B.T. Greenhagen, K. Shi, K. Calabrese, H. Robinson, and J.E. Ladner.
           Structural and functional analysis of the pyocyanin biosynthetic protein phzm from
           pseudomonas aeruginosa. *Biochemistry*, 46(7):1821–1828, 2007.

[Sno]   Snoopy. A tool to design and animate hierarchical graphs. BTU Cottbus, CS Dep.,
           www-dssz.informatik.tu-cottbus.de.

[SR97]   L. F. Shampine and M. W. Reichelt. The MATLAB ODE Suite. *SIAM Journal on
           Scientific Computing*, 18:1–22, 1997.

[Syn08]   SyntheticBiology.org. www.syntheticbiology.org, Cited 20 Jan 2008.

[vHNM+00] J. van Helden, A. Naim, R. Mancuso, M. Eldridge, L. Wernisch, D. Gilbert, and S. J.
           Wodak. Representing and analysing molecular and cellular function in the computer.
           J Biological Chemistry, 9-10(381):921–935, 2000.

[Wil06]   D.J. Wilkinson. *Stochastic Modelling for System Biology*. CRC Press, New York, 1st
           Edition, 2006.

[WWR+98]   B.M. Willardson, J.F. Wilkins, T.A. Rand, J.M. Schupp, K.K. Hill, P. Keim, and
           P.J. Jackson. Development and Testing of a Bacterial Biosensor for Toluene-Based
           Environmental Contaminants. *Appl Environ Microbiol.*, 3(64):1006–1012, 1998.

Image Segmentation by a Network of Cortical Macrocolumns with Learned Connection Weights

Markus Lessmann, Rolf P. Würtz

Abstract Image understanding in the brain or a computer requires segmentation of observed images, i.e., their partition into different semantically-connected parts that each constitute one physical object. This task is fundamental for further processing and analysis of visual information and seems to be accomplished by the brain very easily. Nevertheless it is a very demanding challenge for computer algorithms.

In this article, we present a network of neuronal macrocolumns, which processes contour information by favoring closed contours. The connecting weights have been learned from real image sequences before. Then, segmentation is achieved on the basis of color, texture, and contour information.

1 Introduction

The task of computer vision as well as human vision is to extract semantic information from images or image sequences. This means that the values of a priori unrelated pixels on a camera chip or on the retina of the eye must be organized into larger entities, which can then be identified as objects. A subtask is the division of images into object candidates, a process called segmentation. Psychologically, the properties of this process have been studied for a long time.

Based on experiments in human vision Gestalt psychologists like Max Wertheimer, Wolfgang Köhler and Kurt Koffka formulated eight *Gestalt principles*, formal rules that guide this process [1]. Later this list was extended by Stephen Palmer who

Markus Lessmann
Institute for Neuroinformatics, Ruhr-University Bochum
e-mail: Markus.Lessmann@neuroinformatik.rub.de

Rolf P. Würtz
Institute for Neuroinformatics, Ruhr-University Bochum
e-mail: Rolf.Wuertz@neuroinformatik.rub.de

Please use the following format when citing this chapter:

Lessman, M. and Würtz, R.P., 2008, in IFIP International Federation for Information Processing, Volume 268; *Biologically-Inspired Collaborative Computing*; Mike Hinchey, Anastasia Pagnoni, Franz J. Rammig, Hartmut Schmeck; (Boston: Springer), pp. 177–186.

added 3 further rules [2]. Many of these principles take into account the spatial or temporal context in which different patterns appear.

The details of how the process of segmentation is carried out in the brain, i.e., how networks of neural agents can actually perform this task, are still unknown. On the technical side, the organizational principles that must be applied for meaningful segmentation are intense investigation.

In our project, we use a biological model of basic neuronal ensembles in the brain to perform segmentation by using only non-semantic bottom-up criteria that are easily extractable from input data by application of mathematical filters like Gabor wavelets and Gaussians, both known to be applied by the visual cortex. This means we only rely on a few of the Gestalt rules, namely the *law of proximity*, the *law of similarity*, and the *law of collinearity and curvilinearity*. The first one postulates the vicinity of pixels representing the same object. Distant, unconnected parts of an image are unlikely to belong to the same object. The second rule states the resemblance of areas of the input picture for belonging together. Adjacent points with very similar color and texture are highly likely to belong to the same item. The last rule concerns the contours of an object. In the ideal situation a closed contour is detected by filter functions and separates it from the background. During perception contour segments are grouped by the visual system in a way that smooth curves occur without any acute angles.

To apply these rules contour information is extracted from all 3 channels of colored input images using a biologically-inspired convolution with Gabor wavelets of 8 different orientations, which collect evidence for oriented contour elements, and Gaussians,which calculate mean color values for each pixel. The contour information is preprocessed using non-maximum suppression to keep as much edge information as necessary but delete as many unimportant edges as possible.

The responses of the Gabor and Gaussian filters provide the input to the minicolumns of a macrocolumn network, which will be introduced in the third chapter. The network calculates for each pixel the presence or absence of an edge and, in case of presence, its orientation. This data is combined with information about the similarity of color and texture to decide on the connectedness of neighboring points, finally yielding image segments.

2 Input pictures, filter functions and preprocessing

The program works on color pictures transformed into *CIELab* color space for further processing. This is done because of the resemblance of *Lab* data to the output of retinal neurons and its construction following human perception [3]. By applying Gabor filters to a- and b-channel of an image it is easy to simulate color-coding simple cells of the primary visual cortex that compare red stimuli with green ones and the blue components of the input with yellow ones [4].

Filter functions As mentioned before Gabor wavelets are used throughout this program as well as Gaussians. The Gaussians have a width of 1.0 pixel, the Gabor functions are elliptic with widths $\sigma = 1.25$ and $\tau = 1.0$. Eight equally spaced orientations and five frequencies with a ratio of 1.025. As a model of complex cells in the visual cortex, the *magnitude* of the complex-valued Gabor responses is used.

Preprocessing After the extraction of contour information a lot of the edges perceivable by a human are detected, but also a lot more contours parallel to these meaningful outlines. The reason for this is that a Gabor wavelet placed next to a contour also yields a non-zero-response. To sort these out non-maximum suppression is used. Thus, the orientation of the biggest Gabor filter response at every pixel is determined. Then this biggest value is compared with the best filter responses of pixels on a straight line perpendicular to the orientation of the maximum value, provided that these responses have the same filter alignment. The maximum of these values is labeled by incrementing a counter. If only the pixels with counter values bigger than one are used a lot of unnecessary edges have been erased. The edge elements extracted by the edge detectors and the preprocessing step are shown in figure 1.

3 The Macrocolumn Network

3.1 Fundamentals of the macrocolumn network

The macrocolumn network is based on findings about biological nerve cell clusters in the mammalian cortex and was first introduced in [5, 6]. It consists of a set of macrocolumns each composed of k different minicolumns, which in turn are a bunch of highly interconnected neurons. These minicolumns are represented by their activity $p_\alpha(t)$, which is the fraction of its neurons firing at time t. α is the index of the respective minicolumn. Minicolumns get inputs from the input data and minicolumns of other macrocolumns they are connected with. All minicolumns of the same macrocolumn are inhibited proportional to the maximum of their activity and a modifiable parameter v.

$$\frac{d}{dt}p_\alpha(t) = ap_\alpha(p_\alpha - v \max_{\beta=1..k} p_\beta - p_\alpha^2) \tag{1}$$

The dynamics of the network reveal a bifurcation depending on the value of v. For $v < 0.5$ a macrocolumn can have arbitrarily many active minicolumns, but for $v > 0.5$ only states with at most one active minicolumn are stable. This puts the minicolumns into competition with each other when v is raised from start values below 0.5 to values above 0.5. At the end of this process only the minicolumn with the biggest input can still be active, all others are turned off. That means the system makes a decision on which column received the strongest input.

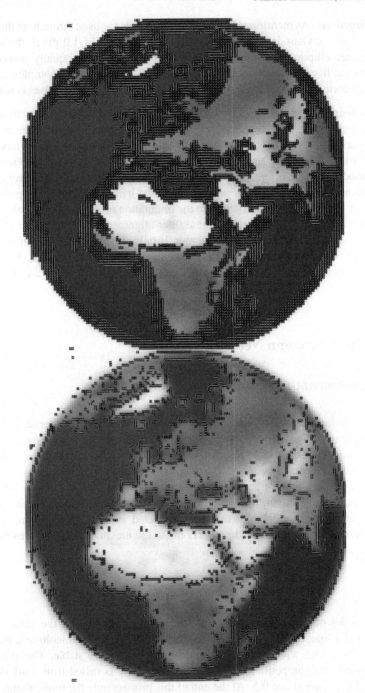

Fig. 1 Preprocessing of a globe image. Top: edges from edge detectors, bottom: edges after non-maximum suppression.

To use information about the local environment the minicolumn i can be connected to other columns in its vicinity. Every connection is defined by the index of the associated minicolumn j and the weight of the connection. These weights are given in a matrix R_{ij}. They are multiplied with the current activity of column j and added to the differential equation.

$$\frac{d}{dt}p_\alpha(t) = ap_\alpha(p_\alpha - v \max_{\beta=1..k} p_\beta - p_\alpha^2) + \kappa \sum_{j=1}^{N} R_{\alpha j} p_j^E + \eta \sigma_t \tag{2}$$

Gaussian noise ($\eta \sigma_t$) is added to distinguish between equally sized inputs. The factor κ defines the relative importance of the connection inputs.

3.2 Use of the macrocolumn network

For the task of contour completion, the macrocolumns must decide whether a pixel of the input picture is part of a contour with a specific orientation. To fulfill this function, preprocessed and normalized responses of Gabor filters constitute the starting values of the minicolumns. Since every picture consists of 3 different color channels and the number of minicolumns should be kept at a minimum due to computational costs one of them is picked to represent this part of the image. This is done by computing the variance of the Gabor filter responses for each channel. That with biggest variance includes the filter response that differs most from responses for other orientations and is highly likely to contain the most distinctive edge. Therefore, its filter values are selected as input to the network. By setting a threshold for the variance, weak edges can be filtered out.

Thus, every minicolumn represents a contour element with a specific orientation. If the minicolumn wins against all others in the macrocolumn then it obtains strong inputs from the filter values and minicolumns of surrounding macrocolumns. That means the picture contains an intense luminance or color contrast at this point or an edge of this orientation fits well to the adjacent contours. An additional 9th minicolumn becomes active if the image does not contain an edge at this pixel. The starting value for this background minicolumn decides how many further contours are detected during the simulation of the network and has to be chosen manually. The higher the start value the smaller is the chance of other minicolumns to compete with the 9th column. It is set to 0.0 for all edges labeled by the preprocessing to be meaningful and to values in the range of 10.0 for all else. That way meaningful contours need not compete with the background and can win easily. All other edges can be suppressed by the background column but still have a chance to win if they get strong inputs via their connections which means they fit well to the surrounding edges.

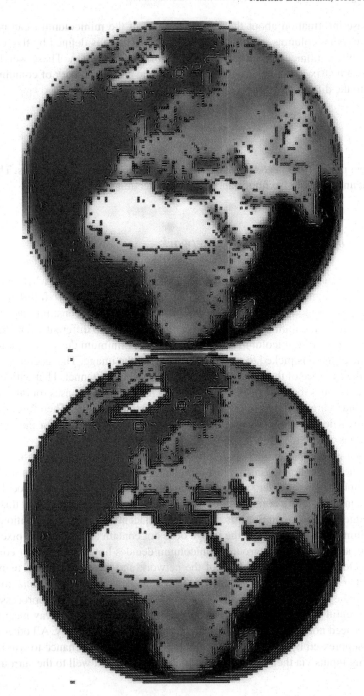

Fig. 2 Processing of a globe image by the macrocolumn network. Top: input to the macrocolumn network, bottom: contours after network processing.

3.3 Learning of the macrocolumn network

In order to enhance contours the network must rely on lateral connections, which support these, even in the case of missing edge segments. So the rule would be that roughly collinear edge detectors should excite each other, while orthogonal ones should inhibit each other. We have made several attempts to specify the connecting weights on a theoretical basis, but the results were not satisfactory. Therefore, we turned to learning them from examples.

3.3.1 Learning rule and normalization

To improve the edge information of input pictures the network has to learn typical spatial relations between different edge orientations from sample pictures (see also [7]). This is done by processing of the sample images with the network and modification of the connection weights using the activities of the minicolumns after each network simulation in conjunction with a Hebbian learning rule as described by Singer [8]:

$$\Delta w_{ij}(t) = \varepsilon \cdot A_i(t) \cdot A_j(t)$$
$$w_{ij}(t) = w_{ij}(t-1) + \Delta w_{ij}(t) \quad \text{if } A_i(t) > \Theta_u \text{ and } A_j > \Theta_u$$
$$w_{ij}(t) = w_{ij}(t-1) - \Delta w_{ij}(t) \quad \text{if } A_i(t) > \Theta_u \text{ and } A_j < \Theta_l$$

In these equations, ε is the learning rate, $w_{ij}(t)$ the connection weight between minicolumns i and j at time t, and Θ_u and Θ_l are appropriately chosen upper and lower thresholds for the activities $A_i(t)$ and $A_j(t)$ of the concerned columns ($\Theta_u = 0.4$ and $\Theta_l = 0.1$). If both minicolumns have activities bigger than Θ_u their correlation is significant and their connection is strengthened. If minicolumn i is very active ($A_i(t) > \Theta_u$) but minicolumn j is not ($A_j(t) < \Theta_l$) it means that j is not of much importance for i and the weight is reduced. In all other cases it is not modified.

Of all connections to the same minicolumn this learning rule favors those with high temporal correlation over those with low correlation. Further competition between them is introduced by normalization over index j. One occurring problem is, that the influence of the weights can completely overcome the filter values of the input image. This is especially unfavorable in the beginning of the learning process, when they are learned from only a few examples and not very reliable. Therefore, they are scaled to 0.1 to keep their influence small enough.

3.3.2 Learning data

Pictures for learning of edge relations should fulfill some criteria:

1. Edges of one single object should be learned to be sure that they have a semantic context and don't belong to neighboring items.

2. The object should be moving so that many possible relations can be observed.
3. The object must not be too inflexible. A very inelastic object like a metal-ball would not reveal many different edge-relations during movement.
4. Only parts of the moving object should be allowed for learning because static background would disturb the statistics of the learning samples.

To match these requirements 2 different sets of training pictures have been chosen. The first one are several photo sequences of a person moving his arm in front of a static background. The other set are the first 8 objects of the COIL100-database [9]. This collection consists of photos of 100 different objects that are rotated about 5° between 2 shots. Therefore they also contain a movement of the item and should be appropriate.

3.3.3 Selection of pixels for learning

During learning, one picture after another of the training set is processed by the network. Then pixels have to be chosen that contain valuable information about edge relations in the training picture. The activities of the macrocolumns at these pixels and of the macrocolumns in an 7×7 environment are read out and used for Hebbian learning. To select the specific pixels a more fundamental segmentation method is used which relies on the movement of objects in the picture (Gestalt law of common fate, see also [7] for discussion). Therefore, movement vectors are calculated for every part of the picture by comparing Gabor responses of a certain pixel in one picture of the sequence with the responses in a 3×3 environment of the current picture element in the following picture. By selecting the pixel with the highest similarity a translation vector can be determined that can be compared with vectors of surrounding pixels. This method does not yield very precise movement information, but it is good enough to separate a moving object from a static background. Each pixel of the area labeled this way containing an edge is chosen for learning purposes.

4 Segmentation

The contours improved by the network are shown in figure 2 and are subsequently used for segmentation with the following procedure.

Color similarity To calculate the similarity of color values of 2 adjacent pixels the difference vector in Lab space and its magnitude are determined. It is one option not to use the L-channel so that only color and no luminance differences are considered. The range from 0 to the maximum difference vector magnitude in the picture is first projected onto the range from $-\frac{\pi}{2}$ to $\frac{\pi}{2}$ and than onto $-\infty$ to ∞ using the tangent. At last the Fermi function $\frac{1}{1+e^{c \cdot y}}$ with $c = 0.1$ transforms the interval into a similarity measure with 0.0 for the maximum difference vector and 1.0 for no distance in color space. Now a threshold decides about the connectedness of 2 neighboring pixels.

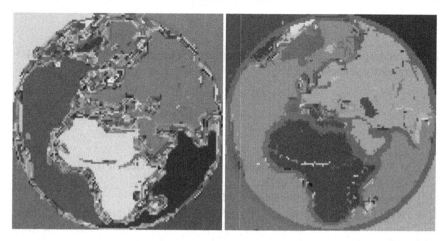

Fig. 3 Segmentation of sample picture before (left) and after processing by the network (right). All adjacent pixels with the same grey value are connected.

Texture similarity Vectors of Gabor responses are now used as texture descriptors, the parameters are as before, but the scales are spread wider. When comparing the similarity of texture features the normalized dot-product of the vectors of Gabor responses is calculated and than mapped onto the range from 0.0 to 1.0 by adding 1.0 and multiplication with 0.5. After doing this the mean value of color and texture similarity is computed and evaluated by means of a threshold.

Integration of contour information Contour information is taken into account by connecting all neighboring pixels containing edges and canceling all connections between edge and non-edge pixels originating from color and texture similarity.

5 Discussion

We have presented a biologically inspired network that organizes image segmentation. This is a collaborative computation, which follows biological inspiration. The required parameters are learned from visual examples using a neurobiologically plausible learning rule. We could show that these learned parameters worked better than the ones we could come up with by theoretical inspection. This is evidence for the correctness of the assumption that the wiring of the visual system reflects the statistics of natural images. The images processed by the contour network were much better suited for segmentation (see figure 3). Full details can be found in [10]. Future developments aim at integrating high-level knowledge of previously seen objects into the segmentation process.

Acknowledgements

Partial funding by the Deutsche Forschungsgemeinschaft (MA 697/5-1, WU 314/5-2) is gratefully acknowledged.

References

[1] M. Wertheimer. Untersuchungen zur Lehre von der Gestalt II. *Psychologische Forschung*, 4:301–350, 1923.

[2] Steven E. Palmer. *Vision Science*. MIT Press, 1999.

[3] David Falk, Dieter Brill, and David Stork. *Seeing the Light: Optics in Nature, Photography, Color, Vision, and Holography*. John Wiley & Sons, New York, 1986.

[4] David H. Hubel and Margaret S. Livingstone. Anatomy and the physiology of a color system in the primate visual cortex. *Journal of Neuroscience*, 4(1):309–356, 1985.

[5] Jörg Lücke, Christoph von der Malsburg, and Rolf P. Würtz. Macrocolumns as decision units. In José R. Dorronsoro, editor, *Artificial Neural Networks – ICANN 2002, Madrid*, volume 2415 of *LNCS*, pages 57–62. Springer, 2002.

[6] J. Lücke and C. von der Malsburg. Rapid processing and unsupervised learning in a model of the cortical macrocolumn. *Neural Computation*, 16(3):501 – 533, 2004.

[7] Carsten Prodöhl, Rolf P. Würtz, and Christoph von der Malsburg. Learning the gestalt rule of collinearity from object motion. *Neural Computation*, 15(8):1865–1896, 2003.

[8] A. Artola, S. Bröcher, and W. Singer. Different voltage-dependent thresholds for inducing long-term depression and long-term potentiation in slices of rat visual cortex. *Nature*, 347:69–72, 1990.

[9] S.A. Nene, S.K. Nayar, and H. Murase. Columbia object image library (COIL-100). Technical Report CUCS-006-96, Columbia University, 1996.

[10] Markus Lessmann. Konturenerkennung mit einem Modell kortikaler Makrokolumnen. Master's thesis, Physics Dept., Univ. of Dortmund, Germany, January 2008.

Integrating Emotional Competence into Man-Machine Collaboration

Natascha Esau, Lisa Kleinjohann and Bernd Kleinjohann

Abstract Emotional competence plays a crucial role in human communication and hence should also be considered for improving cooperation between humans and robots. In this paper we present an architecture and its realization in the robot head MEXI, which is able to recognize emotions of its human counterpart and to react adequately in an emotional way by representing its own emotions in its facial expression and speech output. Furthermore, mechanisms for emotion and drive regulation are presented. Therefor MEXI maintains an internal state made up of (artificial) emotions and drives. This internal state is used to evaluate its perceptions and action alternatives and controls its behavior on the basis of this evaluation. This is a major difference between MEXI and usual goal based agents that rely on a world model to control and plan their actions.

1 Introduction

Quality of human collaboration in accomplishing a specific task does not only depend on their objective competences regarding fulfillment of that task but also on their emotional competence which determines their behavior as a group. The same certainly also holds for tasks to be fulfilled by humans in cooperation with machines or robots. Accordingly, it seems reasonable to equip machines with emotional competence for improving the quality of task accomplishment, i.e. the task's result and the task accomplishing process itself. According to [1] and [2] emotional competence includes several aspects like the abilities to recognize and represent emotions, which are in the focus of many researchers nowadays. But emotional competence also includes mechanisms for emotional behavior, which means adequate reaction on emotions recognized at others, and emotion regulation, i.e. adequate handling of own emotions. In order to show emotional competence in man-machine coopera-

Natascha Esau, Lisa Kleinjohann, Bernd Kleinjohann
C-LAB, University of Paderborn, Germany e-mail: nesau, lisa, bernd@c-lab.de

Please use the following format when citing this chapter:

Esau, N., Kleinjohann, L. and Kleinjohann, B., 2008, in IFIP International Federation for Information Processing, Volume 268; *Biologically-Inspired Collaborative Computing*; Mike Hinchey, Anastasia Pagnoni, Franz J. Rammig, Hartmut Schmeck; (Boston: Springer), pp. 187–198.

tion all four aspects have to be considered as it is realized by the robot head MEXI presented in this paper. MEXI is able to recognize human emotions and to react adequately in its communication behavior (including emotion representation) with regard to emotions exhibited by its human counterpart. Furthermore, MEXI's internal architecture, which is based on emotions and drives for representing its actual state, integrates also the aspect of emotion regulation.

Numerous architectures have been proposed covering one or another aspect of emotional competence. Several expressive face robots have been implemented especially in Japan and USA, where the focus has been on mechanical engineering and design, visual perception, and control. Examples are Saya [3] or K-bot [4] that are constructed to resemble young females and are even equipped with synthetic skin, teeth and hair. They can only recognize and represent emotions. The humanoid robot WE4-RII [5], also determines how stimuli from the environment are evaluated in its current emotional state and how the robot reacts on them. Their model of emotion dynamics is inspired by the motion of a mass-spring system. Arkin et al. [6] discuss how ethological and componential emotion models influence the behavior of Sony's entertainment robots. The homeostasis regulation rules described in [7] is employed for action selection in Sony's AIBO and the humanoid robot SDR as well as in our approach. Canamero and Fredslund [8] realized the humanoid robot Feelix based on an affective activation model that regulates emotions through stimulation levels. Part of it relying on Tomkins' idea that overstimulation causes negative emotions [9] is also adopted in MEXI.

Most similar to our work is the robotic head Kismet built at the MIT [10]. However, KISMET does not recognize emotions like MEXI, but only extracts the intention of its human counterpart e.g. from speech prosody. Although Kismet's motivation system is strongly inspired by various theories of emotions and drives in humans and animals like MEXI, their target is completely different. While KISMET's target is to imitate the development and mechanisms of social interaction in humans, MEXI follows a constructive approach in order to realize the control of purely reactive behavior by its drives and artificial emotions and to show this internal state by corresponding facial expressions and speech utterances.

In this paper we present the robot head MEXI and how its internal architecture (Section 2), which is based on emotions and drives for representing its actual state, integrates the aspects of emotion recognition, representation and regulation (Section 3). Furthermore, we show how MEXI is able to react adequately in its communication behavior with regard to emotions exhibited by its human counterpart (Section 4).

2 Overview of MEXI's Architecture

The robot head MEXI has 15 degrees of freedom (DOF), that are controlled via model craft servo motors and pulse width modulated (PWM) signals. The neck has three DOF (pan, tilt, draw), eyes and ears each have 2 DOF and the mouth has 4

DOF. Furthermore, MEXI is equipped with two cameras, two microphones and a speaker in its mouth for audio output. These facilities allow MEXI to perceive its environment and react on it by representing a variety of emotions like happiness, sadness, or anger via its facial expressions, head movements and by its speech output.

MEXI's software architecture (see Figure 1) is designed according to Nilsson's Triple-Tower Architecture, that distinguishes between perception, model and action tower [11].

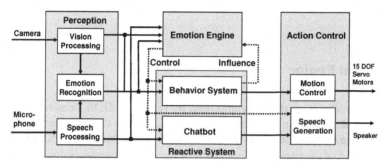

Fig. 1 Architecture of MEXI

The *Perception* component processes MEXI's visual inputs and natural language inputs. The *Vision Preprocessing* is responsible for detecting MEXI's toys or human faces for instance in order to track them with its eyes and head movements. The *Speech Preprocessing* is responsible for speech recognition, which is realized by the commercially available software ViaVoice [12]. One aspect of emotional competence is emotion recognition. In MEXI emotion recognition from human facial expressions [13] as well as from the prosody of human natural speech [14] is supported.

The *Reactive System* allows MEXI to directly respond to its visual and natural speech inputs received from its environment by corresponding head movements, facial expressions and natural speech output. The *Behavior System* is responsible for MEXI's movements. For generating the content of MEXI's answers the slightly extended commercially available *Chatbot* ALICE [15] is used. It receives the textual representation of input sentences from the speech recognition and generates textual output sentences. The content of the sentences generated by the Chatbot is influenced by MEXI's *Emotion Engine*. If MEXI is happy for instance, by chance corresponding output sentences like "I am happy." are generated.

MEXI uses its current perceptions and its internal state, representing the strength of its emotions and drives, to determine its actions following two principle objectives: One is to feel positive emotions and to avoid negative ones for itself and also for its human counterpart. The second objective is to keep its drives at a comfortable (homeostatic) level. In a feedback loop MEXI's internal state and its current perceptions are used by the *Emotion Engine* to configure the *Behavior System* in such a way that appropriate behaviors are selected in order to meet the two objectives stated above (see Section 3).

The *Action Control* component controls the servo motors for the above mentioned 15 DOF via the component *Motion Control* and is responsible for generating natural speech outputs with prosodic features corresponding to MEXI's current emotional state via the component *Speech Generation*. Together with the *Behavior System* the *Action Control* is responsible for the *emotion representation* aspect of emotional competence. The remainder of the paper will concentrate on the realization of emotion regulation mechanisms and emotional behavior by the *Emotion Engine*.

3 Emotion Engine

The *Emotion Engine* is the key component for realizing MEXI's emotional competence regarding emotion regulation and adequate behavior, which results in corresponding emotional expressions in MEXI's face and speech output. Since MEXI's main purpose of cooperation is communicating with its human counterpart the basic behaviors provided by the *Behavior System* were selected accordingly. However, in different contexts (and with according actors) additional behaviors realizing other tasks could be integrated as well. As already mentioned, MEXI's cooperation is determined by the two objectives, to keep its drives at a comfortable (homeostatic) level and to feel positive emotions and to avoid negative ones (for itself and its human counterpart). These objectives determine MEXI's actions resulting in a pro-active regulation of emotions and drives. Furthermore, an intrinsic self-regulation mechanism for automatic decay of emotions and cyclical increase/decrease of drives, as it can be observed also in humans, is realized. These regulation mechanisms and the representation of MEXI's emotions and drives are described next. Afterwards the feedback loop between *Behavior System* and *Emotion Engine* realizing MEXI's emotional behavior is explained.

3.1 Emotion, Drives and their Regulation

For MEXI we distinguish a small set of basic emotions that can be represented easily by MEXI's facial expression and audio output, i.e. anger, happiness, sadness and fear. Happiness is a positive emotion, which MEXI strives for, while the others are negative ones that MEXI tries to avoid. In order to realize a simple control mechanism for MEXI we consider mainly cyclical homeostatic drives. Homeostatic drives motivate behavior in order to reach a certain level of homeostasis. Examples in humans are hunger or thirst. MEXI has for instance a communication drive or a playing drive. Violation of the homeostasis is the more likely to cause according behavior the more the discrepancy between homeostasis and the actual "value" increases.

Furthermore, an exploratory drive lets MEXI look around for new impressions, if none of the homeostatic drives causes any behavior.

In MEXI's *Emotion Engine* each emotion e_i is represented by a strength value between 0 and 1 and each drive d_i by a strength value ranging from -1 to 1. For each emotion a threshold **th** defines when MEXI's behavior will be configured to show this emotion, e. g. by corresponding facial expressions (see Figure 2, shaded areas). This is done by increasing the gain values of the respective desired behavior(s) (see next subsection). Drives have an upper and lower threshold **th$_u$** and **th$_l$** that define when a drive strives to dominate MEXI's behavior (shaded areas, see Figure 2) by increasing the respective gains. This also avoids that MEXI's behavior oscillates in order to satisfy competing drives. In order to realize pro-active behavior MEXI's drives increase and decrease in a cyclical manner. Between its thresholds a drive is in homeostasis.

The course of a drive $d_i(t)$ and an emotion $e_i(t)$ over time t is determined by the following equations, where $\Delta d_i(t)$ and $\Delta e_i(t)$ denote their change between time points $t-1$ and t:

$$d_i(t) = d_i(t-1) + \Delta d_i(t), \text{ with } \Delta d_i(t) = c_{d_i} \cdot \delta_{d_i}(t) \cdot k_{d_i}(t), \tag{1}$$

$$e_i(t) = e_i(t-1) + \Delta e_i(t), \text{ with } \Delta e_i(t) = c_{e_i} \cdot k_{e_i}(t). \tag{2}$$

c_{d_i} and c_{e_i} are positive values from the interval $]0, 0.5]$ that determine the gradient of a drive or emotion due to intrinsic regulation. They were determined experimentally. This intrinsic regulation happens even, if MEXI receives no perceptions or executes no behavior that could influence the respective drive d_i or emotion e_i. The acceleration factors $k_{d_i}(t)$ and $k_{e_i}(t)$ may accelerate or slow down the intrinsic regulation if their absolute value is > 1 (acceleration) or < 1 (slow down). They determine the influence of external stimuli or of MEXI's own behavior regarding a specific drive or emotion. How these factors are determined is described more closely below.

For drives an additional factor $\delta_{d_i}(t) \in \{-1, 1\}$ determines the direction of their course. It is dependent on the previous values $\Delta d_i(t-1)$ and $d_i(t-1)$. When a drive for example becomes 1 at the time t ($d_i(t) = 1$), it starts decreasing due to MEXI's internal regulation mechanisms and $\delta_{d_i}(t+1)$ becomes -1 meaning that the drive is going to be satisfied. When a drive becomes -1 ($d_i(t) = -1$), it automatically starts increasing and $\delta_{d_i}(t+1)$ becomes 1. This factor allows to realize a cyclical behavior of drives. For emotions this factor can be ommitted, since here only an automatic decrease is realized.

The principle development of emotions and drives over time is shown in Figure 2. Also the principle determination of acceleration factors due to external stimuli is explained below.

The solid curve in Figure 2 a) shows the development of a drive d_i. In order to realize pro-active behavior MEXI's drives increase and decrease in a cyclical manner. As a default excitation MEXI's drives would follow a sine wave (dashed curve) and incorporate only internal regulation mechanisms. This is realized by assigning

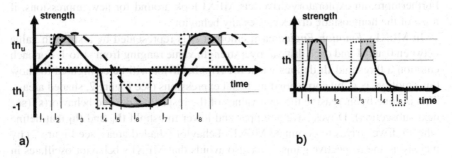

Fig. 2 Emotions and Drives

δ_{d_i} the values 1 and - 1 alternatingly. For drives the interval $]th_l,th_u[$ represents the level of homeostasis, 1 represents a very large drive striving for its satisfaction and -1 expresses that the drive was overly satisfied. Stimuli, i.e. perceptions and own behavior, influencing the drive are depicted as dotted line in Figure 2. Stimuli may accelerate a drive's increase (intervals I_2, I_8). The stimuli that satisfy the drive cause a steeper decrease (interval I_4). In these intervals ($t \in \{I_2, I_4, I_8\}$) the acceleration factor is greater than 1 ($k_{d_i}(t) > 1$) and accelerates a drive's increase (I_2, I_8) or its decrease (I_4). Other stimuli may cause a slower increase or decrease of the drive. In this case the factor $k_{d_i}(t)$ has a value between 0 and 1 ($0 < k_{d_i}(t) < 1$). The slower increase is shown in interval I_6, where a negative stimulus indicates its over-satisfaction. In some timing intervals MEXI's perceptions or their absence do not influence the depicted drive (indicated by a zero line of the stimuli, e.g. interval $I_1, I_3, I_7, ...$ In these cases the factor $k_{d_i}(t)$ is equal to 1 and the course of a drive runs in parallel with the excitation function.

Imagine for example that MEXI sees a human face in a state where its communication drive d_{com} is decreasing ($\delta_{com} = -1$). Since the current perception signals a potential communication partner for MEXI, its communication drive should be satisfied faster. This is reached by setting the acceleration factor $k_{com}(t)$ to a value $k > 1$. If no stimuli, that may influence the communication drive, are recognized $k_{com}(t)$ remains 1 and does not accelerate the normal internal regulation of the communication drive. If for instance the person disappears (the face becomes smaller) the communication drive might decrease slower, and hence the value of $k_{com}(t)$ should be in the interval $]0,1[$.

Figure 2 b) shows the development of a positive emotion like happiness over time as the solid curve. The dotted curve shows the duration and evaluation of MEXI's current percepts. For positively evaluated percepts the curve is above the time axis, for negative ones it is below the time axis. In contrast to drives, for each emotion only one threshold **th** defines when MEXI's behavior will be configured to show this emotion, e. g. by corresponding facial expressions (shaded area) (see next subsection).

The acceleration factor $k_{e_i}(t)$ for an emotion e_i depends on the previous value of the emotion $e_i(t-1)$, the evaluation of the current perception and also on the current drive state. If a drive d_i concerning an emotion e_i increases, MEXI reacts in a neutral way, i. e. $k_{e_i}(t) = 0$ and hence also $e_i(t) = 0$. If that drive starts decreasing ($\delta_i(t)$ changes its value from 1 to -1 and hence Δd_i becomes < 0) then the emotion e_i increases very rapidly and $k_{e_i}(t) \geq 1$. The decrease of an emotion could also be accelerated by setting $k_{e_i}(t)$ to a value $k \leq -1$.

The increase of a positive emotion may be caused by positive perceptions of its environment (intervals I_1, I_3) and by the drives (their fulfillment). Hence, for $t \in \{I_1, I_3\}$ the acceleration factor $k_{e_i}(t)$ is larger than 1 ($k_{e_i}(t) > 1$). The decrease happens automatically with a certain adaptable amount per time unit, if the positive stimulus has disappeared (intervals I_2, I_4). In this case the acceleration factor $k_{e_i}(t)$ is equal to -1. By a negative stimulus the decrease of $e_i(t)$ is accelerated by setting $k_{e_i}(t) < -1$ (interval I_5).

Imagine for instance the communication drive d_{com} and its impact on the emotion happiness e_{happy}. Assume that it is not satisfied ($d_{com} > th_u$ and $\Delta d_{com} < 0$) initially. Then happiness may be increased by a positive perception perhaps a human face. This should result in an accelerated increase of e_{happy}, which is reached by setting the respective acceleration factor $k_{happy} > 1$. If the face suddenly disappears, happiness decreases ($k_{happy} = -1$). If according stimuli are recognized, happiness could decrease even faster resulting in $k_{happy}(t) < -1$.

3.2 Control of the Behavior System by the Emotion Engine

The feedback loop between *Behavior System* and *Emotion Engine* realizing MEXI's behavior as adequate reactions to its environment and the emotions of its human counterpart is depicted in Figure 3. For clarity reasons only some behaviors, drives and emotions are shown.

For realizing MEXI's *Behavior System* the paradigm of *Behavior Based Programming* developed by Arkin [6] is applied. So called basic behaviors b like *Smile, Look around, Follow face, Avert gaze* or *Sulk*, which may be either cooperative or competitive, are mixed by an accumulator Σ to compute the nominal vector $\mathbf{R}$ for the actor system from the fixed-sized vectors $\mathbf{C_b}$. A $\mathbf{C_b}$ generated by the basic behavior b contains 15 triplets $(c_{s,b}, v_{s,b}, m_{s,b})$ (one for each servo motor s) consisting of the nominal value $c_{s,b}$, a vote $v_{s,b}$ for each of the nominal values and a mode flag $m_{s,b}$ (cooperative vs. competitive).

Using the vote values a behavior can signal whether the output is of high importance ($v_{s,b} = 1$) or should have no influence at all ($v_{s,b} = 0$). Via this mechanism the influence of MEXI's external percepts, e.g. the emotions recognized at its human counterpart, on its behavior is realized. Apart from that also its internal emotional state may influence the actual behavior by setting appropriate *gain* values. The accumulator combines external votes and internal gains of a behavior b by calculating a weight $w_{s,b} = g_{s,b} \cdot v_{s,b}$ as their product. This weight is used to do the ranking of

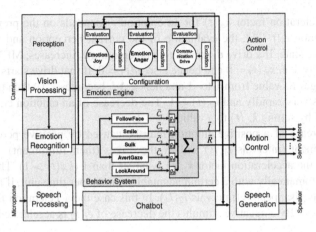

Fig. 3 Detailed Architecture

the behaviors. If the behavior b with highest weight is competitive, a winner-takes-all strategy is used and the resulting nominal value for servo s is $r_s = c_{s,b}$. Else a weighted median of all cooperative values is calculated. This calculation as well as the calculation of the Influence vector $\mathbf{I}$ is described e.g. in [16].

The *Excitation* components allow to realize MEXI's regulation mechanisms as described above. The *Configuration* component determines from the actual percepts (including the emotional state of MEXI's counterpart) and the internal state of emotions and drives which behaviors of the *Behavior System* should be preferred by setting the gain values for each behavior b. If a drive increases the gain for the behaviors, that will satisfy that drive, is increased by a certain (variable) amount per time. If a drive decreases the gains for the "satisfying" behaviors are decreased respectively. If a percept causes a certain emotion in MEXI the emotion strength is increased. If a certain threshold is reached the gains for according behaviors e. g. those that generate the corresponding facial expression are set to 1 and the Chatbot and Speech Generation are instructed to produce according speech output and prosody. Conflicts between drives may be solved for instance by a fixed priority. An example showing these interdependencies is described in Section 4. The gain calculations are presented elsewhere [16].

The *Evaluation* component evaluates the influence vector $\mathbf{I}$ calculated by the accumulator in the *Behavior System* in order to determine whether the preferred behavior really dominates the actual actions activated by the *Motion Control*. This decision is positive, if for competitive behaviors the influence equals 1 and for cooperative behaviors a specific threshold is exceeded. If the behavior was initiated by an emotion the negative excitation function is switched on for subsequent cycles in order to let the emotion strength decrease automatically (k_{e_i} becomes negative). In case of drives the strength of the drive is decreased by a specific amount (δ_{d_i} becomes -1). This is repeated, if the behaviors are preferred also for subsequent cycles until the lower threshold of the drive is reached.

4 Example Session

For an example, how MEXI reacts adequately to human emotions, have a look at the diagrams in Figure 4 that show how MEXI behaves when at time t_0 a person appears and MEXI sees a human face. At t_1 MEXI recognizes that the person has a sad facial expression and tries to distract her from that feeling by playing. - MEXI plays by getting shown its toy and tracking it. - At t_3 MEXI detects the toy and plays with it until it becomes boring and MEXI looks at the human face again at t_4.

The upper two diagrams show the votes $v_{s_h,b}$ and $v_{s_m,b}$ for two groups of servo motors involved in the execution of different sets of behaviors. Motors for head and eye movements $s_h \in \{1, \dots 7\}$ are for instance needed for the behaviors *FollowFace, FollowToy* or *LookAround* which are competitive, but not for *Smile* or *Sulk* that are cooperative behaviors. Vice versa, motors for movements of the mouth corners $s_m \in \{8, \dots 11\}$ are needed for *Smile* or *Sulk* and not for the other behaviors. For behaviors not involving a group of servos the corresponding votes are constantly set to zero. The votes of the other behaviors depend on MEXI's current perceptions and somehow reflect its current action tendency due to these perceptions not taking into account its internal state.

The votes $v_{s_h,b}$ for the behavior $b = FollowFace$ and for $b = FollowToy$ are set to one when MEXI sees a human face (t_0 and afterwards) or its toy respectively (t_3 and afterwards). The behavior *LookAround* which lets MEXI look for interesting things (e. g. human faces) corresponds to MEXI's exploratory drive. It describes a kind of default behavior for the head movement motors. Therefore, the votes $v_{s_h,b}$ for $b = LookAround$ are constantly set to 0.5 and the gain is set to 1.0.

The behavior *Smile* only involves the motors for mouth movements. Since MEXI by default should be friendly the corresponding votes $v_{s_m,b}$ for $b = Smile$ are set to a high value of 0.8, when MEXI perceives a human face (starting at t_0). When MEXI classifies the human face as sad at t_1, it adapts its own facial expression to a neutral expression and no longer intends to smile. Hence, the votes $v_{s_m,b}$ for *Smile* are set to zero and remain there until MEXI sees its toy at t_3 and wants to smile again due to this perception ($v_{s_m,b} = 0.8$ at t_3).

The gain values reflect MEXI's action tendency due to its current internal state. The weight combines gains and votes, i.e. internal and external action tendencies and the influence shows which behavior is really executed (and to which extent in the case of cooperative behaviors). The influence of *LookAround* is one until t_0 and MEXI is looking around since nothing else is seen. When MEXI's communication drive d_{com} increases also the gain for *FollowFace* g_{FF} rises until d_{com} reaches its upper threshold ($th_u = 1$) and also g_{FF} is set to one at t_0. Since MEXI sees a face at t_0, also the weight of *FollowFace* becomes one and *FollowFace* is executed (its influence is one) because its weight is now higher than that of *LookAround*. At t_1 MEXI detects the sad face of its human counterpart and stops smiling (vote of *Smile* is set to zero, hence weight and influence become zero too). Since MEXI now wants to distract the human from her sadness by playing with its toy, its own *PlayToy* drive increases to one causing the gain for *FollowToy* to increase as well. Since MEXI now wants to play, the communication drive starts decreasing faster

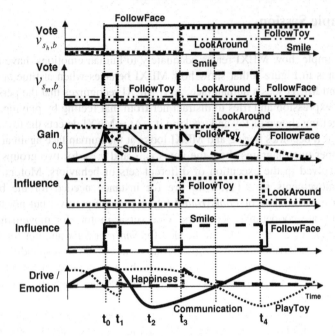

Fig. 4 Example Session

until it becomes zero at t_2. Now the weight of *LookAround* is larger than that of *FollowFace* and MEXI looks around (influence of *LookAround* = 1 from t_2 to t_3) until at t_3 the weigth of *FollowToy* is larger than that of *LookAround*, since MEXI detects its toy at t_3 and the corresponding vote $v_{s_h,FollowToy}$ becomes one. Then also the *PlayToy* drive and due to this also the gain of *FollowToy* start decreasing until the drive is minus one ($th_l = -1$) at t_4. Since at t_4 the communication drive has reached its upper threshold ($th_u = 1$), the gain of *FollowFace* is set to one and its weight is the highest again resulting in execution of *FollowFace* after t_4 (influence = 1).

The emotion *Happiness* is set to one each time the influence of *FollowFace* or *FollowToy* starts rising (t_0,t_3), i.e. the communication drive or the playing drive starts being satisfied (δ_{com} is set to -1). As result of the rising of the emotion *Happiness* (threshold = 1) the gain value of the behavior *Smile* also is set to one and MEXI starts smiling, because also the vote is one (t_0 and t_3). *Happiness* automatically decreases caused by the respective excitation function (e.g. from t_3 until t_4). This decrease becomes faster by setting $k_{happy} < -1$, when MEXI receives corresponding percepts like a sad human face (t_1). MEXI smiles until the weight of *Smile* becomes zero because of MEXI's percept ($v_{s_m,Smile} = 0$ at t_1 and t_4). Another reason to stop smiling may be that the gain of *Smile* has decreased to zero because of a corresponding decrease of *Happiness* due to its excitation function e_{happy}.

5 Conclusion

In this paper we presented the architecture of the robot head MEXI and how it supports emotional competence in human-robot cooperation. The paper concentrates on two important aspects of emotional competence: how internal regulation mechanisms for emotions and drives can be realized and how adequate behaviors for external emotion regulation as well as for reaction on human emotions can be incorporated. We presented MEXI's software architecture and how it is used to realize its actions without any explicit world model and goal representation. Instead MEXI's artificial emotions and drives maintained by the Emotion Engine are used to evaluate its percepts and control its future actions in a feedback loop. The underlying Behavior System and the Emotion Engine are based on the behavior based programming paradigm extending Arkin's motor schemes to a multidimensional model of reactive control. This architecture supports a constructive approach for synthesizing and representing MEXI's artificial emotions and drives rather than emulating human ways of "feeling". MEXI integrates also components for emotion recognition components from facial expression and natural speech. Based on these building blocks MEXI can cooperate with humans in real-time by communication behavior, which could be extended to other tasks by integrating appropriate actor facilities and basic behaviors into the Behavior System and the Emotion Engine. Our experiences with MEXI on different public exhibitions and fairs show that MEXI, although realizing only a restricted set of emotions and drives, attracts human spectators and maintains their communication interest. However, it remains a question for psychological studies, how different humans react when they cooperate with emotionally extended machines. We plan to investigate how the constructive approach of emotions and drives for behavior control can be transferred to other application domains than human-robot communication.

References

1. P. Salovey, J. D. Mayer, *Emotional intelligence. Imagination, Cognition, and Personality*, 9, 185-211.
2. W. Seidel, *Emotionale Kompetenz. Gehirnforschung und Lebenskunst*, SPEKTRUM AKADEMISCHER VERLAG 2004.
3. H. Kobayashi, F. Hara, A. Tange, *A basic study on dynamic control of facial expressions for face robot*, Proceedings of the International Workshop on Robots and Human Communication, 1994.
4. Robots are getting more sociable, MSNBC News, 2003, http://www.msnbc.msn.com/id/3078973/
5. A. Takanishi, H. Miwa, *Emotional control of emotion expression humanoid robot WE-4RII*, Workshop on Building Humanoid Robots (Humanoids 2004), Los Angeles CA, USA, November 2004.
6. R. Arkin, M. Fujita, T. Takagi, T. Hasekaawa, *An ethological and emotional basis for human-robot interaction*, Robotics and Autonomous Systems 42 (2003) 191-201.

7. R. Arkin, *Homeostatic Control for a Mobile Robot, Dynamic Replanning in Hazardous Environments*, Proceedings SPIE Conference on Mobile Robots, Cambrige, MAA, pp. 240-249, 1988.
8. L. Canamero, J. Fredslund, *I show you how I like you - can you read it in my face?*, IEEE Transactions on Systems, Man and Cybernetics 31 (5), 2001.
9. S. S. Tomkins, *Affect theory*, in Approaches to Emotion, K. R. Scherer and P. Ekman (eds.), Hillsdale, NJ: L. Erlbaum, pp. 163-195, 1984.
10. C. Breazeal, *Affective Interaction between Humans and Robots*, in J. Kelemen and P. Sosk (editors), Proceedings of ECAL 01, Prague, pp. 582-591, Springer 2001.
11. N. J. Nilsson, *Artificial Intelligence - A New Synthesis*, Morgan Kaufmann Publishers, 1998.
12. Via Voice, http://www-306.ibm.com/software/voice/viavoice/
13. N. Esau, E. Wetzel, L. Kleinjohann, B. Kleinjohann, *Real-Time Facial Expression Recognition Using a Fuzzy Emotion Model*, Proceedings of the IEEE International Conference on Fuzzy Systems (FUZZ-IEEE 2007), London, UK, July 2007.
14. A. Austermann, N. Esau, B. Kleinjohann, L. Kleinjohann, *Prosody Based Emotion Recognition for MEXI*, Proceedings of the IEEE/RSJ Int. Conference on Intelligent Robots and Systems (IROS 2005), Edmonton, Canada, August 2005.
15. Alicebot, http://www.alicebot.org/
16. N. Esau, L. Kleinjohann, B. Kleinjohann, *Integration of Emotional Reactions on Human Facial Expressions into the Robot Head MEXI*. In: Proceedings of the IEEE/RSJ International Conference on Intelligent Robots and Systems (IEEE/RSJ IROS 2007), San Diego, USA, 29. Okt. - 2. Nov., 2007.

Self-optimized Routing in a Network on-a-Chip

Wolfgang Trumler, Sebastian Schlingmann, Theo Ungerer, Jun Ho Bahn, Nader Bagherzadeh

Abstract Many-cores are on the cusp of becoming state-of-the-art processor technology for the next decade. To guarantee efficient communication between multiple cores, a Network-on-a-Chip (NoC) is considered as an alternative to overcome the limitations of the ubiquitous bus technology.

In this paper, we present an approach to further improve the routing in an NoC with a self-optimized routing strategy. We extended the routers of a network to measure their load and to send an appropriate load information to their direct neighbors. The load information is used to decide in which direction a packet should be routed to avoid hot-spots. Evaluation results show a significant increase in the network throughput. With the self-optimized routing, the NoC is capable of routing up to two times more packets compared to the original routing algorithm proposed by Lee and Bagherzadeh, 2006.

1 Introduction

In 2007 Intel announced a prototype 80-core tera-scale processor [11] using Network-on-a-Chip (NoC) [7] technology. NoC is used as an alternative to the ubiquitous bus technology in order to facilitate communication among many cores. As the process technology shrinks and more cores are integrated on the same chip, the current bus approach for communication among cores will not be sufficient and a technology such as NoC is needed.

Wolfgang Trumler, Sebastian Schlingmann, Theo Ungerer
Department of Computer Science, University of Augsburg, Eichleitnerstr. 30, 86159 Augsburg, Germany, e-mail: {trumler, schlingmann, ungerer}@informatik.uni-augsburg.de

Jun Ho Bahn, Nader Bagherzadeh
Department of Electrical Engineering and Computer Science, University of California, Irvine, California, USA, e-mail: {jbahn, nader}@uci.edu

Please use the following format when citing this chapter:

Trumler, W., Schlingmann, S., Ungerer, T., Bahn, J.H. and Bagherzadeh, N., 2008, in IFIP International Federation for Information Processing, Volume 268; *Biologically-Inspired Collaborative Computing*; Mike Hinchey, Anastasia Pagnoni, Franz J. Rammig, Hartmut Schmeck; (Boston: Springer), pp. 199–212.

In an NoC system, processor cores exchange messages using a network as transportation system that is constructed from multiple point-to-point data links interconnected by routers such that messages can be relayed from any source module to any destination module over several links by making routing decisions at the local routers. NoCs apply message passing communication networks similar to massively parallel systems. For NoCs, the advantage of a low latency communication, compared to the off chip communication on high-speed channels among processors, offers new possibilities but also new challenges for the routing in such networks.

There is only limited room for improvements to the topology of an NoC compared to the 2D-mesh due to space and energy constraints on a chip. Therefore, most NoCs employ a simple 2D-mesh for communication infrastructure. On the other hand, significant efforts have been spent on the optimization of routing for NoCs concerning both, the overall throughput and the average latency.

The O1TURN algorithm [18] for example has a provable near-optimal worst case throughput. Another routing algorithm with good performance and deadlock free routing is ROMM [14].

In this paper, we present an approach to increase the network throughput and to lower the average latency based on the local load information of the nodes. The nodes exchange their local load values with their neighboring nodes, which route incoming packets based on this information. The basic idea of this self-organizing, adaptive routing algorithm is inspired by the self-optimization algorithm [20] for load balancing in large scale networks, which is based on the notion of the human hormone system. The underlying architecture [13] for our algorithm, which does not rely on virtual channels, has been developed at the University of California in Irvine.

The artificial hormone system described in [20] piggy backs load information on the outgoing messages. This information is extracted by the receiving node, which decides wether to transfer load, in form of a service, to the origin of the message. In this scenario the communication was constrained by the uniquely identified communication partners of each service. We assumed that one service will not send the message to all other services, but only one of its communication partners at a time. This simulates the way information flows in object oriented software, where one object can call methods of only a few other objects. Furthermore, it is similar to the way hormones distribute their information to only those parts of the tissue which have receptors for these specific hormones and thus can act on this information.

The communication pattern of the self-optimizing algorithm in this paper is more related to the way information is distributed known from the process of morphogenese [21], first described by Turing in 1952. Turing mathematically described his idea of messengers that diffuse into neighboring regions in the tissue of animals and plants, used to organize the creation of regular structures. The concentrations of different messengers are responsible for the creation of the patterns of a zebras or the regular leaf structure of woodruff for example.

The communication pattern of the NoC routers does not change over time nor does the structure or the layout of the routers on the die change. In this sense the self-optimization is more related to the findings of Turing. On the other hand, the simple approach of local load values, without the complicated differential equations of the morphogeneses, is more related to the artificial hormone system as desribed in [20].

The remainder of this paper is structured as follows. The next section describes the architecture of the underlaying hardware. In Section 3, the calculation of the local load is described and the routing algorithm is explained in detail. Simulation results are presented in Section 4 and related work is discussed in Section 5. The paper closes with a conclusion and the description of future work in Section 6.

2 Basic Router Architecture

The network topology of the NoC is a 2D-mesh with *NxM* routers. Figure 1 shows the architecture of a single router. The router consists of three subsections, the left, the right, and the internal router.

The task of the internal router is to inject packets into the network. Packets that arrived at their destination are also ejected by the internal router. The left router is used to route packets to the left (west) and the right router routes packets to the right (east), respectively. Both routers can route packets to the north as well as to the south but there is no connection between the east and west direction. This assumption divides the network into two separate networks where packets can go either to the east or the west of a router, but there is no turning back in the horizontal routing direction. This architectural approach guarantees the NoC to be dead-lock free [5].

Clock boosting was introduced to improve the performance of the NoC. A packet consisting of multiple flits (flow control digits) can be transmitted at different clock speeds. With clock boosting only the routing decision for the head flit is done at normal clock speed. After the route for the packet is chosen, the body flits can be routed with an increased clock speed. By multiplying the clock frequency more than one body flit can be routed during a normal clock cycle. The clock boosting can double or quadruple the basic clock frequency, allowing two or four body flits to be routed at the same time.

The routing decision of a router is straight forward using a clockwise priority scheme to select the packets for routing. Starting from the north (top) input channel the router examines the head flit from the buffer and tries to set the route for this packet if possible. If there is no head flit in front of the buffer or if the route can not be set because the output channel is already occupied by another packet, the router picks the next buffer in a clockwise order and repeats the aforementioned steps. More details about the clock boosting and the routing decision can be found in [13].

The internal FSM of the router is easy to implement but does not take the current load situation of the network into account. Our approach is to improve the routing decision by considering the current load of the neighboring routers, finding routes to avoid hot-spots. Furthermore, when the original algorithm stops routing due to a congested network, our algorithm can use alternative routes, bypassing the heavily loaded routers, which leads to a higher overall network throughput.

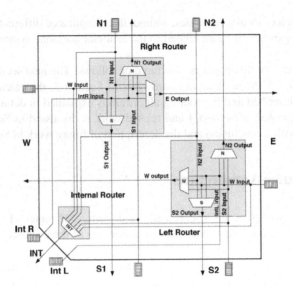

Fig. 1 Architecture of a router

3 Routing Algorithm

Every router performs a routing decision whenever a packet arrives. The routing decision consists of two consecutive steps. First, a *routing function* creates a set of possible output channels (next destinations) for a packet. Afterwards, the *selection function* calculates quality values for all possible routes and selects the most appropriate one for the routing of the packet.

In the current setup, the routing function creates a set with all nodes that lead to the desired direction of the packet. For packets going from west to east (right direction) the set may contain the north, west, and south channels. For packets going from east to west (left direction) possible output channels can be north, east, and south, respectively. If a packet has to be routed in south or north direction only, this desired direction is used for the routing.

The local load of a router is propagated to its neighbors so they can decide if the router is a good choice to route packet. The local load is not used for the local routing decisions but it is crucial for the routing decision of the neighboring routers. The propagation of the load values is described in Section 3.2.

3.1 Self-Optimization-Algorithm

3.1.1 Selection Function

The Self-Optimization-algorithm calculates the quality of a possible route based on Equation 1. To yield a load value in the range of 0 to 100 all parameter values are normalized.

$$quality = direction - \%of\ remaining\ flits - 4 * pload \qquad (1)$$

The first value (*direction*) is a sort of bonus if the route leads to the destination of the packet. A value of 200 is added for a route heading towards the destination and 0 for a route which leads the packet farther from its target.

The value *%of remaining flits* expresses the percentage of a packet that must still be sent on the selected output channel. If there is currently a route set for a packet, the amount of the remaining flits of the packet is divided by the packet length. This value is used to penalize a route, if there is already a packet on that route. The amount of remaining flits is used to give an idea of how long the route will be occupied.

The *pload* value is the load value of the neighboring router in the desired direction. The calculation of the *pload* value is described in Section 3.2. The propagated load of the considered direction (*pload*) is multiplied by four and subtracted from the other values. The higher the load of the neighboring routers, the less attractive is the route in that direction.

3.1.2 Calculating the Load of a Router

The SO-algorithm tries to avoid hot-spots by spreading the offered load as good as possible to the available routers as long as free capacity is available. The load value, which is also in the range from 0 to 100, is used for the calculation of a channel's quality. The load calculation of the SO-algorithm uses the utilization of the buffers as a degree for the load of a router. The more the buffers are filled with flits, the higher is the load of a router. Therefore, the load can be calculated by the fraction of currently available flits for the available buffer size.

$$load = \frac{local}{maxload} * 100 \qquad (2)$$

The value *local* is the sum of the flits in the input buffers of a router. The amount of flits from injection buffers are limited by the capacity of an input buffer, because the injection buffers are assumed to be unlimited for simulation purposes. The maximum utilization of a router, *maxload*, is reached if all input buffers are completely filled with flits.

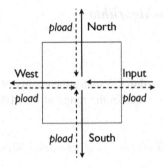

Fig. 2 Propagation scheme for the left router

3.2 Load propagation

To propagate the local load, a router sends the load value to its neighboring routers. Therefore, the overall load of a router is calculated and then propagated to the neighbors.

The propagated load value, *pload*, not only includes the local load, but also the load values of the surrounding routers. The information of the surrounding routers is used to increase or decrease the local load of a router. This guarantees that the load information of a heavily loaded router is not only sent to its direct neighbors but is spread from the center of a possible hot-spot to the surrounding parts of the network. The closer a router is located to a hot-spot the higher is the propagated load value, which is used to decide if a packet should be routed to this node or not.

The calculation of the value *pload* is shown in Equation 3. Two-thirds of *pload* are taken from the local load value *load*. One third of the propagated load depends on the load of the neighboring routers.

$$pload = \frac{1}{3}\left(2*load + \frac{\sum pload(dir)}{|pload(dir)|}\right) \tag{3}$$

The load that is propagated into one direction depends on the load of the routers from the possible direction that can be used to route a packet. For example, the load of the left router (see Figure 2) , the router that can route either to the north, south, or west, takes the load values from these three neighboring routers and sends this information to the right (east) router. Depending on the direction, the *pload* value incorporates the load value of up to three neighboring routers. For the *pload* value propagated to the south of the right router only the load information from the north and west router is taken into account, because the routers can not route a packet back to the direction where it came from. Thus, the only possible routes are to the north and west.

4 Evaluations

We conducted extensive experiments with different network sizes ranging from 4x4 up to 16x16. Next, we will describe the traffic patterns used for the generation of the network load and discuss the results for a 4x4 and 8x8 2D-mesh network in comparison to the results of the design from the University of California in Irvine. Afterwards, we compare the performance gain of the network in terms of increased network throughput.

4.1 Traffic generation

We evaluated our algorithms with four different traffic patterns as proposed in [7] (Chapter 9). We used the same four traffic patterns as they did at the UCI to have a basis for a direct comparison of the results.

The traffic patterns used are *matrix transpose, bit reverse, bit complement*, and *uniform random*. The target address of a packet is generated out of the source address of a node by applying one of the aforementioned traffic patterns. For these simulations, the width and height of the 2D-mesh network is assumed to be a power of two. Without this assumption some of the traffic patterns might produce invalid destination addresses.

For every simulation setup, we conducted multiple runs and calculated the average network latency to minimize the impact of possibly good circumstances in one simulation run. The network had a warm-up phase of 1000 cycles at the beginning of every simulation. The measurement was done during 100,000 cycles following the warm-up phase. The simulation stopped if the average delay of the packets exceeded a threshold of 200 cycles.

The injection of the flits was chosen to be the worst case, which means, that all nodes inject their flits at the same time resulting in a high network load. Assuming that the nodes would distribute the flit injection over time, the results are even better than shown in the charts.

The packet injection rate can be calculated from the injection rate of the flits. In our simulations, we used a fixed packet length of nine flits (one head flit and eight body flits). Therefore, the packet injection rate can be calculated by dividing the flit injection rate by nine.

4.2 4x4 NoC

The first simulations were done on a 4x4 2D-mesh with the aforementioned traffic patterns. The results are shown in Figure 3. The charts show the generated traffic on the x-axis and the measured average latency on the y-axis. The generated traffic is given in flits per node per cycle which is a value ranging from 0.05 to 1. If the generated traffic is 1, a new flit is injected into the internal buffer of the routers at every node in every cycle. If the traffic is below 1, there are a few cycles delay between the injection of new flits, e.g. at 0.5 a new flit is injected every second cycle.

For a direct comparison of the results from Irvine and our results, we plotted all the data into one chart. Every chart shows the results with 1x, 2x, and 4x clock boosting for both.

Concerning the matrix transpose traffic pattern there is hardly any improvement of the SO-algorithm compared with the original algorithm. Both can route all packets with a nearly constant delay up to the maximum load, if the clock boosting is used. The saturation point of the network seems to be about the same for both algorithms.

The bit reverse traffic pattern first shows considerable differences. The saturation point for the SO-algorithm is about 0.65 and about 0.45 for the original algorithm without clock boosting. The maximum throughput of the SO-algorithm is even better for the 2x clock boosting. While the original algorithm begins to saturate at about 0.8, the SO-algorithm routes all packets with nearly the same delay up to the maximum injection rate. The results with 4x clock boosting are again the same for both algorithms.

Bit complement traffic shows interesting results in terms of the algorithms saturation points. The original algorithm has a saturation point of 0.35, 0.5, and 0.85, respectively. The SO-algorithm first increases the latency, which creates a kind of plateau phase. This plateau has a different length for different clock boosting values. The plateau also appears at the 4x clock boosting, but the SO-algorithm can route all the packets with only a slight latency increase up to the maximum injection rate and does not reach the saturation point.

The uniform random traffic pattern seems to be the hardest of all four traffic patterns, because the saturation point of the original algorithm is reached earliest. The SO-algorithm shows very good behavior especially for the 2x and 4x clock boosting. In both cases, the SO-algorithm does not reach the saturation point but can route all packets up to the maximum injection rate. The uniform random traffic pattern shows similar behavior for the SO-algorithm than in the previous traffic pattern. There are two plateaus for the 1x and 2x clock boosting at different injection rates. The first plateau is also visible for the 4x clock boosting but the second one does not appear.

The explanation for the appearance of the plateaus can be derived from the way the SO-algorithm selects the routes for the destination of a packet. As long as there are enough good alternative routes to the destination, the SO-algorithm automatically selects the next best (shortest) path. For example, a packet at the node $(1,1)$ should be routed to $(3,2)$ and if the buffers of the router to the east $(2,1)$ are filled, the SO-algorithm routes the packet to $(1,2)$ instead. If the injection rate increases, more buffers are completely filled and more congestions arise in the network. At the same time, since the quality of the alternative routers are also getting worse, the SO-algorithm selects routes by avoiding both shortest but heavily loaded routes.

If the SO-algorithm starts routing packets not on one of the shortest paths, the average latency increases due to the additional hops and the possible congestions.

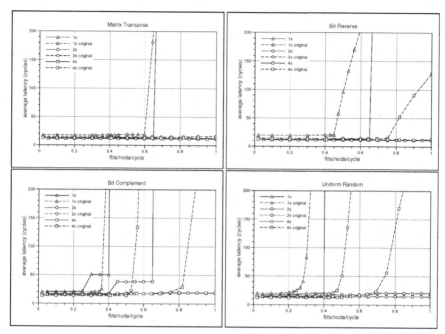

Fig. 3 Simulation results for a 4x4 2D-mesh

On the other hand, the alternative routes around heavily loaded nodes give the SO-algorithm the ability to defer the ultimate saturation point to a much higher injection rate and in most cases it can handle the maximum injection rate.

4.3 8x8 NoC

We also conducted experiments with 8x8 2D-mesh networks and the four traffic patterns to see how the SO-algorithm performs in larger networks.

The amount of injected flits depends on the amount of nodes. Therefore, if the number of nodes is doubled in each dimension, the amount of nodes in the network increases exponentially and so does the amount of flits injected into the network. On the other hand, the exponential growth in the network size might offer more alternative routes for the SO-algorithm to be selected as routes for the packets.

The results of the simulations of an 8x8 2D-mesh are shown in Figure 4 for 1x, 2x, and 4x clock boosting. The original algorithm reaches its saturation point for all traffic pattern at an injection rate considerably below the maximum injection rate.

With the matrix transpose traffic pattern, the SO-algorithm performs much better than the original algorithm. Especially with 4x clock boosting, where the original algorithm saturates at about 0.65, the SO-algorithm can route packets up to the maximum injection rate with a low average latency. The same applies for the bit

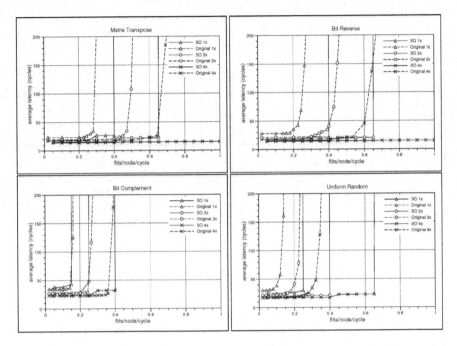

Fig. 4 Simulation results for an 8x8 2D-mesh

reverse traffic pattern where the original algorithm performs even worse while the SO-algorithm performs slightly better than with the matrix transpose traffic pattern.

The bit complement seems to be challenging for both, the original and the SO-algorithm. In this case the SO-algorithm does hardly outperform the original algorithm, in contrast to the other three traffic patterns. We had similar results for the 4x4 network. One of the influencing factors seems to be the network architecture, which is chosen to avoid dead-locks by design. Because of the two separated networks for the horizontal directions, the SO-algorithm does not offer enough choices for alternative routes. A packet cannot be routed back in horizontal direction in contrast to the vertical direction where a packet can go back and forth if the packet has been routed horizontally, at least once after a vertical transfer.

In the lower right chart of Figure 4 the results of the uniform random traffic pattern are depicted. In this case, the SO-algorithm can handle the double injection rate than the original algorithm before it reaches the saturation point. The difference with the bit complement traffic pattern is due to better distribution of the traffic over the whole network, which seems to be the favored kind of traffic for the SO-algorithm.

The plateaus known from the 4x4 network can be observed at all four charts of 8x8 network. As already mentioned, this behavior occurs when the SO-algorithm starts to use alternative routes, which are not on the shortest path to the destination.

4.4 Performance comparison

To compare the network throughput of the original and the SO-algorithm Tables 1 and 2 show the throughput gain in percentage for the 4x4 and 8x8 network, respectively. The original algorithm's results are taken as base for the calculation of the throughput gain. Therefore, a value of 0% means that both algorithms perform equally and a value of 100% means that the SO-algorithm performs twice as good as the original algorithm.

For an algorithm that reaches its saturation point, an average injection rate latency of 50 cycles was taken for the base calculation. When the original algorithm saturates and the SO-algorithm does not, the throughput gain becomes much higher than the calculated value shown in the tables. This is because in all these cases the average latency of the SO-algorithm is much better than 50 cycles. These values are marked with a * to denote that the original algorithm saturated in contrast to the SO-algorithm.

Table 1 Throughput gain for a 4x4 NoC

Traffic Pattern	1x	2x	4x
Matrix Transpose	13%	0%	0%
Bit Reverse	41%	* 23%	0%
Bit Complement	8%	75%	* 20%
Unifrom Random	37%	* 96%	* 35%

The throughput gain for the 4x4 network ranges from 0% for the matrix transpose to more than 96% for the uniform random traffic pattern. As mentioned before in the case of the uniform random traffic pattern the throughput gain is even higher due to the fact that the original algorithm has a 50 cycles delay for the chosen injection rate while the SO-algorithm has an average delay of about 21 cycles. For the 8x8 2D-mesh network, the throughput gain ranges from 0% for the bit complement up to 127% for the uniform random traffic pattern.

Table 2 Throughput gain for a 8x8 NoC

Traffic Pattern	1x	2x	4x
Matrix Transpose	37%	35%	* 53%
Bit Reverse	73%	58%	* 63%
Bit Complement	7%	0%	8%
Unifrom Random	127%	90%	103%

The tables confirm our initial assumption that the SO-algorithm will perform better for larger networks due to the larger amount of possibles routes. The bit complement is the sole exception to this assumption. Further investigations are needed to better understand the reasons for the poor throughput gain in this case.

5 Related Work

In designing Network-on-a-Chip (NoC) systems, there are several issues to be considered, such as topology, routing algorithm, performance, latency, and complexity. Because of its flexibility, architectures based on NoC are getting more attention. As a feasible topology in NoC systems, the mesh is getting popular for its modularity; it can be easily expanded by adding new nodes and links without any modification of the existing node structure.

Another issue in NoC environment is the routing algorithm. In terms of delivering mechanism, wormhole routing has increasingly been advocated as a method of reducing message routing latency. In wormhole routing, a packet is decomposed into flits or flow control units, and the packet follows through the network one flit after another. On the other hand, in terms of the way of selecting a path among the sets of possible paths from source to destination, the routing algorithms are classified as deterministic/oblivious and adaptive ones [5]. The oblivious/deterministic routing algorithms choose a route without considering any information about the network's present condition, resulting in relatively simple design complexity. Adaptive routing algorithms use the state of the network such as the status of a node or link, the status of buffers for network resources, or history of channel load information. Even though the adaptive routing algorithms utilize the flexibility in routing paths, the hardware design complexity is usually increased. Depending on the degree of adaptivity, minimal adaptive and fully adaptive routing algorithms are refined. DOR (dimension-ordered routing) [19], ROMM [14], and O1TURN [18] are examples of deterministic or oblivious algorithms. Some researchers have developed better performance routing algorithms using adaptive routing algorithms [10, 4, 6, 2, 9, 8]. The SO-algorithm is an adaptive routing algorithm that does not use virtual channels.

The adoption of virtual channel (abbreviated to VC) has been prevailing because of its versatility. By adding virtual channels and proper utilization of their channels, deadlock-freedom can be easily accomplished. Network throughput can be increased by dividing the buffer storage associated with each network channel into several virtual channels. By proper control of virtual channels, network flow control can be easily implemented [3]. Also to increase the fault tolerance in a network, the concept of virtual channel has been utilized [1, 12]. However, in order to maximize its utilization, allocation of virtual channels is a critical issue in designing routing algorithms [22, 16]. Furthermore, the buffers of the virtual channels are very expensive in terms of chip size. The extra chip size needed for the additional logic of our SO-algorithm is negligible compared to the chip size needed for the buffers of virtual channels and the logic for the channel allocation.

A similar approach, using stress values, is described in [15]. The stress values are exchanged between the direct neighbors in the network. With our calculations the load value is not only exchanged with the direct neighbors, but diffuses to the surrounding area of load value's source. Furthermore, there are less choices for our algorithm to chose the best routes due to the two separated networks, which guarantee deadlock free routing by design.

6 Conclusion and Future Work

In this paper we presented a routing algorithm for an Network-on-a-Chip that yields a significant throughput gain for 2D-mesh networks. The SO-algorithm calculates the load of a router based on the amount of flits in the buffers. With this approach the network throughput can be significantly increased. The throughput gain is up to 127% compared to the original algorithm. The throughput gain is highest for the uniform random traffic pattern, which in our opinion is especially relevant for our current research where we will investigate task allocation mechanisms on an NoC. The expected traffic in such a dynamic and steadily changing environment will lead to a traffic pattern comparable to the uniform random.

Based on our latest results, we have plenty of ideas to improve the SO-algorithm. The first will be to combine the SO-algorithm with another approaches that is described in [17]. The idea is to use the SO-algorithm as a base value for the local load and to increase or decrease it by the amount of I/O operations a router can process during a cycle. If the input channels of a router are filled with flits, the router is loaded to a level of 100% (concerning the SO-algorithm) but the situation is aggravated if the router is blocked and cannot perform any read or write operations. On the other hand, if the router is full but can transfer a maximum number of flits, it might be a better choice.

References

1. S. Chalasani and R. V. Boppana. Fault-tolerant wormhole routing algorithms for mesh networks. *IEEE Trans. Comput.*, 44(7):848–864, 1995.
2. G.-M. Chiu. The odd-even turn model for adaptive routing. *IEEE Trans. Parallel Distrib. Syst.*, 11(7):729–738, 2000.
3. W. J. Dally. Virtual-channel flow control. In *17th annual international symposium on Computer Architecture (ISCA '90)*, pages 60–68, New York, NY, USA, 1990. ACM.
4. W. J. Dally and C. L. Seitz. Deadlock-free message routing in multiprocessor interconnection networks. *IEEE Trans. Comput.*, 36(5):547–553, 1987.
5. W. J. Dally and B. Towles. *Principles and Practices of Interconnection Networks*. Morgan Kaufmann Publishers, 2004.
6. J. Duato. A new theory of deadlock-free adaptive routing in wormhole networks. *IEEE Trans. Parallel Distrib. Syst.*, 4(12):1320–1331, 1993.
7. J. Duato, S. Yalamanchili, and L. Ni. *Interconnection Networks – An Engineering Approach*. Morgan Kaufmann Publishers, 2003.
8. C. J. Glass and L. M. Ni. Maximally fully adaptive routing in 2D meshes. In *Proceedings of the 1992 International Conference on Parallel Processing*, volume I, Architecture, pages I:101–104, Boca Raton, Florida, 1992. CRC Press.

9. C. J. Glass and L. M. Ni. The turn model for adaptive routing. *J. ACM*, 41(5):874–902, 1994.
10. J. Hu and R. Marculescu. Dyad - smart routing for network-on-chip. In *41-st Annual Conf. on Design and Automation*, pages 260–263, 2004.
11. Intel Corporation. Intel's teraflops research chip. http://download.intel.com/research/platform/ terascale/teraflops/ FINAL_TeraflopsResearchChip_Overview.pdf, November 2007.
12. F. Jipeng Zhou; Lau. Adaptive fault-tolerant wormhole routing with two virtual channels in 2d meshes. *Parallel Architectures, Algorithms and Networks, 2004. Proceedings. 7th International Symposium on*, pages 142–148, 10-12 May 2004.
13. S. E. Lee and N. Bagherzadeh. Increasing the throughput of an adaptive router in network-on-chip (noc). In *3rd International Conference on Hardware-Software Codesign and System Synthesis (CODES+ISSS)*, Seoul, Korea, Oktober 22-25 2006. ACM.
14. T. Nesson and S. L. Johnsson. ROMM routing on mesh and torus networks. In *Proc. 7th Annual ACM Symposium on Parallel Algorithms and Architectures SPAA'95*, pages 275–287, Santa Barbara, California, 1995.
15. E. Nilsson, M. Millberg, J. Oberg, and A. Jantsch. Load distribution with the proximity congestion awareness in a network on chip. In *DATE '03: Proceedings of the conference on Design, Automation and Test in Europe*, pages 11126–11127, Washington, DC, USA, March 2003. IEEE Computer Society.
16. H. Rezazad, M.; Sarbazi-azad. The effect of virtual channel organization on the performance of interconnection networks. *Parallel and Distributed Processing Symposium, 2005. Proceedings. 19th IEEE International*, pages 8 pp.–, 4-8 April 2005.
17. S. Schlingmann. Selbstoptimierendes routing in einem network-on-a-chip. Master's thesis, University of Augsburg, September 2007.
18. D. Seo, A. Ali, W.-T. Lim, and N. Rafique. Near-optimal worst-case throughput routing for two-dimensional mesh networks. In *32nd International Symposium on Computer Architecture, 2005. ISCA '05*, pages 432–443, Madison, Wisconsin USA, 4-8 June 2005.
19. H. Sullivan and T. R. Bashkow. A large scale, homogeneous, fully distributed parallel machine, i. In *ISCA '77: Proceedings of the 4th annual symposium on Computer architecture*, pages 105–117, New York, NY, USA, 1977. ACM.
20. W. Trumler, T. Thiemann, and T. Ungerer. An artificial hormone system for self-organization of networked nodes. In *IFIP Conference on Biologically Inspired Cooperative Computing*, pages 85–94, Santiago de Chile, August 2006. Springer-Verlag.
21. A. M. Turing. The chemical basis of morphogenesis. *Philosophical Transactions of the Royal Society of London. Series B, Biological Sciences*, 237(641):37–72, August 1952.
22. A. S. Vaidya, A. Sivasubramaniam, and C. R. Das. Impact of virtual channels and adaptive routing on application performance. *IEEE Trans. Parallel Distrib. Syst.*, 12(2):223–237, 2001.

On Robust Evolution of Digital Hardware

Tobias Knieper, Bertrand Defo, Paul Kaufmann, and Marco Platzner

Abstract In this paper we investigate whether multi-objective evolution of digital hardware components has advantages over single-objective evolution in terms of convergence and robustness. To that end, we experimentally compare a standard genetic algorithm to several multi-objective optimizers on a set of test problems. The results show that, for more complex test problems, the multi-objective optimizers TSPEA2 and NSGAII indeed outperform the single-objective genetic algorithm as they more often evolve correct circuits, and mostly with less computational effort.

1 Introduction

Self-adaptive and self-optimizing systems are able to react to changes in the environment and the internal system state autonomously. Systems with such self-X properties find applications in, for example, highly complex scenarios where classical methods fail, or in scenarios which require autonomous operation for long mission periods. To design such systems, often principles from biology or sociology are transferred into engineering domains and combined with modern hardware and software technology to form what is denoted as *organic computing*.

In our work, we focus on organic computing methods to develop hardware components. More than a decade ago, the emergence of reconfigurable hardware architectures together with natural computing methods gave rise to the field of *biologically-inspired hardware,* which includes several areas [1]: *Evolvable hardware* denotes the combination of evolutionary algorithms with reconfigurable hardware to construct self-adaptive and self-optimizing hardware systems. *Embryonics* tries to apply developmental processes as found in multicellular organisms to design fault-tolerant circuits with self-repair and self-healing capabilities. *Immunotronics* uses principles of the immune system to support fault tolerance and protection for

University of Paderborn
e-mail: {tknieper, bertrand, paul.kaufmann, platzner}@upb.de

Please use the following format when citing this chapter:

Knieper, T., Defo, B., Kaufmann P. and Platzner, M., 2008, in IFIP International Federation for Information Processing, Volume 268; *Biologically-Inspired Collaborative Computing*; Mike Hinchey, Anastasia Pagnoni, Franz J. Rammig, Hartmut Schmeck; (Boston: Springer), pp. 213–222.

hardware circuits. Finally, *neural hardware* denotes hardware implementations of models of the nervous system.

We concentrate on evolvable hardware, a term coined by de Garis [2] and Higuchi [3] in 1993. The common denominator of all evolvable hardware approaches is the application of evolutionary algorithms directly at the hardware level. Here, hardware means both digital and analog electronic circuits, and the hardware level comprises all models of hardware, from configuration bitstreams for reprogrammable devices over netlists of gates to behavioral descriptions. In a sense, evolutionary algorithms exploit a form of collaborative computing as they keep a set of individuals and try to improve them over generations by applying biologically-inspired operators such as selection, mutation, and crossover. Especially crossover allows to pass on information between individuals.

Most work in evolvable hardware has been focusing on evolving functionally good or correct components. In contrast, we consider several objectives and include also the required hardware area and the resulting circuit speed in the evolution-ary process. While a few previous approaches applied two-stage fitness functions [4, 5], we employ multi-objective evolutionary algorithms (MOEAs). There are two motivations for using MOEAs in evolvable hardware design: First, MOEAs are de-signed to keep diversity in the population and generate Pareto sets of circuits. Au-tonomous systems can switch between the solutions in the Pareto set in order to react on quickly changing resource situations and performance goals [6]. Second, MOEAs can possibly be used for a faster and more robust evolution of functionally good circuits. The argument is that putting too much selection pressure on only one objective (the functional quality of the circuit), instead of keeping the population di-verse with respect to other objectives (area and speed of a circuit), one might more easily get stuck in the optimization process and, hence, need a higher computational effort to evolve components with acceptable functional quality.

In this paper, we want to investigate whether multi-objective evolution of dig-ital hardware components has advantages over single-objective evolution in terms of convergence and robustness. To that end, we experimentally compare a standard genetic algorithm (GA) to several recent multi-objective evolutionary algorithms on a set of test problems. The paper is structured as follows: In Section 2, we present the hardware representation model that is used to encode circuit individuals, and the computation of objectives. The different evolutionary algorithms including four multi-objective optimizers are discussed in Section 3. Section 4 shows the test prob-lems, the experimental setup and the results, before Section 5 concludes the paper.

2 Hardware Representation Model and Metrics

Cartesian genetic programming (CGP) is a very popular hardware representation model introduced in [7]. CGP is a structural hardware model, where a circuit is formed by combinational logic blocks arranged in a two-dimensional array and an interconnect (wires) between the blocks. Figure 1 presents the CGP model and its parameters. The array consists of $n_c \times n_r$ combinational blocks, n_i primary inputs, and n_o primary outputs. The primary inputs can be connected to the inputs of any logic block in the array. A logic block in column j has n_n inputs that can be connected to the columns $j - l, \ldots, j - 1$ of the array and to the primary inputs, respectively. This ensures that no combinational feedback loops are generated. A combinational block implements one out of n_f different logic functions of its inputs.

An individual is defined by its chromosome (genotype). The length of the chromosome is given by $n_c \cdot n_r (n_n + 1) + n_o$. Each of the logic blocks in the array is defined by $n_n + 1$ values, one for each input and one for the logic function. Additionally, an n_o-tuple of values selects the block outputs that are connected to the primary outputs of the array.

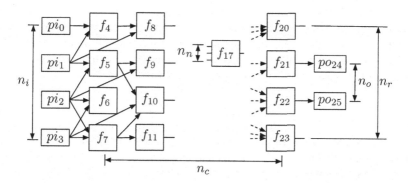

Fig. 1 The cartesian genetic programming model for hardware representation with its main parameters

The main reason for the popularity of the CGP model is its closeness to the architectures of field-programmable reconfigurable hardware arrays (e.g., FPGAs or coarse-granular arrays). The block functions can be set to simple two-input gates, to n_n-input lookup tables, or to more complex word-based arithmetic operators. The interconnect can model bit wires or busses. While the original CGP model implicitly encodes block placement, more recent CGP variants rely on only one row of blocks, i.e., $n_r = 1$ and $l = n_c$. Routing is not encoded in the CGP model, mainly to keep the genotype (chromosome length) short and, thus, to increase the efficiency of the evolutionary operators.

Generally, the genotype has to be mapped to a corresponding phenotype for evaluating the fitness. The phenotype represents the actual circuit and is achieved from the genotype by removing all blocks of the array that do not contribute to the outputs. Note that there might still be redundancy in the phenotype. An important previous result with the CGP model is that propagating redundant and currently unused structures inside the chromosomes through the search process of the evolutionary algorithm can increase the speed of convergence [7].

In this paper, we are interested in evaluating the circuits' fitness with regard to three objectives: the functional quality, the speed of the circuit, and the required hardware area. Accordingly, we have to define three metrics to evaluate circuit fitness. Following related work in evolvable hardware, we use logic and arithmetic functions as test problems [8]. As we know the correct outputs for all input value combinations for these functions, we determine the *functional quality* as reciprocal of the summarized square error distances between the output vectors of an evolved individual c and a correct function c^*:

$$f(c) = \frac{1}{1 + \frac{1}{N} \sum_{i=1}^{N} ham(c^*(i), c(i))^2}, \tag{1}$$

where N denotes the number of test vectors and *ham* refers to the Hamming distance of two bit vectors. A correct circuit receives a functional quality of one.

We estimate the delay of a circuit by the number of wires or logic blocks on the longest path. Given the CGP model, the delay is in the range $\{0, \ldots, n_c + 1\}$. A delay of zero means that the longest path of the circuit connects an input directly with an output. A delay of n_c means that the longest path traverses all logic blocks of the model, whereas a delay of $n_c + 1$ indicates that none of the outputs is connected to an input. The fitness with respect to circuit *speed* is determined as:

$$speed(c) = 1 - \frac{delay(c)}{n_c + 1} \tag{2}$$

The speed metrics equals one for the fastest possible circuit (a circuit that maps primary inputs directly to primary outputs) and zero for a circuit that has no connection at all from primary inputs to primary outputs.

The number of logic blocks used by a circuit c, denoted as *used_blocks*(c), is in the range $\{0, \ldots, n_c \cdot n_r\}$. Based on this value, we define a circuit's fitness with respect to *area* as:

$$area(c) = 1 - \frac{used\_blocks(c)}{n_c \cdot n_r} \tag{3}$$

A circuit of minimal size, i.e., a circuit not using any logic block, receives an area of one, a circuit that utilizes all available logic blocks has an area of zero.

3 Multi-objective Optimizers

In this section, we review the multi-objective evolutionary optimizers SPEA2, TS-PEA2, NSGAII, and μGA that are compared in our experiments. As a reference algorithm, we use a standard single-objective genetic algorithm (*GA*). The parameters for GA are set as follows: The top 5% of the individuals are selected and transferred without any modification to the next generation. Then, we apply two-stage binary tournament as selection scheme, followed by a two-point crossover with a recombination probability of 90%, and mutation. We choose the mutation rate such that only one combinational block or wire is mutated each time the mutation operator is applied. Each recombined child is mutated exactly once.

SPEA2 is a recent multi-objective evolutionary optimizer introduced by Zitzler et al. [9]. SPEA2 maintains two sets of individuals: an archive that contains non-dominated individuals and a breeding population. In each generation, the two sets are merged and the fitness of the individuals is evaluated. The non-dominated individuals are then copied to the new archive. If the archive exceeds a predefined maximum size, SPEA2 applies a nearest neighbor density estimation technique to thin out clusters on the Pareto front. The fitness assigned to an individual considers thenumber of individuals it dominates (the dominance count), the number of individuals that are dominators (the dominance rank), and a density estimate based on the k-th nearest neighbor method. All individuals undergo a binary tournament selection which selects parents for recombination and mutation.

TSPEA2 is an algorithm we have devised in order to increase selection pressure on one objective while trying to keep diversity [6]. This should be beneficial for evolving circuits with a correctness property, where we will not be satisfied with a circuit unless the functional quality reaches a predefined level. Both SPEA2 and TSPEA2 use an archive and a breeding population and a selection scheme based on Pareto dominance ranking. TSPEA2, however, checks as a first selection rule in a binary tournament whether one of the two individuals dominates the other regarding the main objective. TSPEA2 has been motivated by an earlier algorithm, MO-Turtle GA presented by Trefzer et al. [10], that preferred a main objective over several other objectives during the evolution of analog circuits.

NSGAII was presented by Deb et al. in [11]. NSGAII separates the population into a hierarchy of Pareto fronts. The first level Pareto front is formed by the non-dominated individuals. These individuals are then removed from the population, and the second level Pareto front is formed by the now non-dominated individuals, and so on. A new elite population is filled by incrementally adding these Pareto fronts, starting with the level one front. In case the addition of the next level Pareto front exceeds the population's capacity, a density metric is used to select among the individuals of that front. A breeding population is created by using a standard GA scheme. Here, the selection operator takes the hierarchical Pareto front information and the density metric into account to achieve both diversity and a minimal distance to the optimal Pareto front.

μGA follows the original idea of Goldberg [12] who observed that a small number of individuals in a population is often sufficient for a converging optimization process. Consequently, he suggested an optimization scheme where a GA operates on a very small population. The situation in which all individuals have similar chromosomes is called nominal convergence. If such a nominal convergence is reached, the search process is relaxed by inserting randomly initialized individuals into the population. In [13], Coello Coello and Pulido combined the idea of Goldberg with the Pareto front diversity technique of Knowles and Corne [14]. Their μGA algorithm relies on three populations: an external archive population which contains non-dominated individuals of high diversity, the population memory which corresponds to the classical GA breeding population, and a non-replaceable population which carries arbitrarily initialized individuals for the case of nominal convergence. In each step, a standard GA is applied on a small set of randomly selected individuals from the breeding and the non-replaceable population. After reaching nominal convergence, the best individuals are copied to the breeding and the external population. After several iterations of this scheme, a part of the breeding population is replaced by non-dominated individuals from the external population.

4 Experiments and Results

We have applied the different evolutionary optimizers to the following commonly used benchmarks for evolving digital circuits [15, 16, 5]: the 6 and 7 even parity function, $2+2$ and $3+3$ adders, and 2×2 and 3×3 multipliers. For the experiments, we have configured the CGP model as a single line of two-input gates (nodes). For the 6-parity function, the chromosome consists of 12 nodes, for the 7-parity function of 15 nodes, for the $2+2$ adder and 2×2 multiplier of 50 nodes, and for the $3+3$ adder and 3×3 multiplier of 200 nodes. As for the reference GA, the MOEAs rely on a two-point crossover with a recombination probability of 0.9. In each new individual, a single gene is mutated by modifying either the logic function or an input connection. The function set for the nodes is not restricted for the adder and multiplier experiments, i.e., the node function can be an arbitrary function of two inputs. For the parity experiments, however, the node function set has been restricted to AND, NAND, OR and NOR. Particularly, the XOR logic function is excluded, as otherwise the evolution of correct parity functions is not a challenge. All experiments have been conducted using the MOVES framework [17] for multi-objective evolutionary optimization of digital circuits.

As an example result, Figure 2 displays the development of the average functional quality for the 2×2 multiplier circuit. For this test problem, TSPEA2 shows the fastest convergence, followed by NSGAII, GA, SPEA2, and μGA. This result clearly shows that some multi-objective optimizers outperform the standard single-objective GA in evolving functionally correct circuits.

As we are interested in the asymptotical behavior of the algorithms regarding the functional quality of the evolved circuits, we have conducted several optimization runs for each test problem. We have stopped an optimization run when a correct circuit has been evolved. Otherwise, we have stopped the evolution after a predefined number of fitness evaluations. For the parity function this limit has been set to $14 \cdot 10^6$ fitness evaluations, for the 3×3 multiplier to $6 \cdot 10^6$ fitness evaluations, and for all other experiments to $20 \cdot 10^6$ fitness evaluations.

We use two metrics to compare the algorithms. The first is the number of successfully evolved circuits among all runs of an experiment. This metric relates to robustness. The second metric is the computational effort as defined by Koza in [18] and can only be determined if a sufficient number of experiment runs produces correct circuits. In each run the optimization goal, i.e., the evolution of a functionally correct circuit, will be reached by some generation i. Having M fitness evaluations

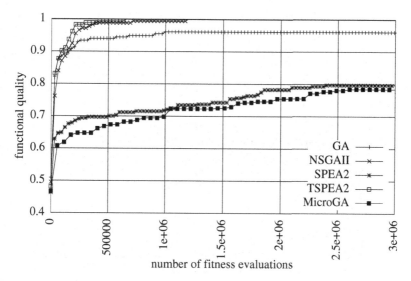

Fig. 2 Average functional quality of the best individuals over the number of fitness evaluations for the 2×2 multiplier (10 experiments runs)

per generation, the probability of reaching the optimization goal by generation i can then be expressed as follows:

$$P(M,i) = (\text{\#succeeded runs by generation } i)/(\text{\#runs})$$

From that we can determine $R(z)$, the number of independent runs that have to be conducted to reach the optimization goal with a certain probability z:

$$R(z) = \lceil \log(1-z)/\log(1-P(M,i)) \rceil$$

The estimated overall number of fitness evaluations required to reach the goal with probability z is then set to:

$$I(M,I,z) = M \cdot (i+1) \cdot R(z)$$

For each experiment with given M and z, the minimal value for $I(M,i,z)$ is determined as the computational effort of the experiment. In our experiments, we have set z to 99%.

The complete set of results is presented in Table 1. This table shows the computational effort and the number of successfully evolved circuits for 10 experiment runs for each test problem. SPEA2 and μGA did not succeed in evolving a sufficient number of correct circuits within the predefined number of fitness evaluations. Therefore, we did not compute the computational effort for these optimizers. The ranking of the algorithms with respect to the computational effort is shown in Table

2, where bold values indicate that the optimizers were able to evolve a functionally correct circuit in each single experiment run.

Table 1 Computational effort and number of correctly evolved circuits for standard GA and the MOEAs. The computational effort is given in multiples of 10^6. SPEA2 and μGA could not evolve a sufficient number of correct circuits to determine the computational effort.

	6-parity	7-parity	2 + 2 add	3 + 3 add	2 × 2 mult	3 × 3 mult
GA	0.09 / 10	0.25 / 10	0.09 / 10	6.63 / 9	0.79 / 8	– / 5
TSPEA2	0.15 / 10	2.02 / 10	1.42 / 10	1.55 / 8	0.59 / 10	1.89 / 10
NSGAII	1.14 / 10	3.65 / 10	1.10 / 10	3.61 / 10	1.04 / 10	3.29 / 9
SPEA2	– / 2	– / 0	– / 1	– / 0	– / 0	– / 0
μGA	– / 1	– / 0	– / 6	– / 2	– / 7	– / 0

From the results, we observe that the simpler functions, i.e., parity and $2 + 2$ adder, are easily evolved by the GA, and also by TSPEA2 and NSGAII. However, the multi-objective optimizers TSPEA2 and NSGAII require considerably more effort to evolve correct circuits. For the $3 + 3$ adder and the multipliers, the GA could not compete with TSPEA2 and NSGAII either in computational effort, the number of successfully evolved circuits, or both. The results indicate that with rising benchmark complexity, evolving a diverse population with regard to objectives such as circuit speed and area yields an improved robustness.

Table 2 Computational effort ranking. Bold values indicate experiments where every run produced a functionally correct circuit.

	6-parity	7-parity	2 + 2 add	3 + 3 add	2 × 2 mult	3 × 3 mult
GA	**1**	**1**	**1**	3	2	3
TSPEA2	**2**	**2**	3	**1**	**1**	**1**
NSGAII	**3**	**3**	2	2	**3**	2
SPEA2	4	4	5	5	5	4
μGA	5	4	4	4	4	4

5 Conclusion

In this paper, we have presented an experimental comparison of several multi-objective evolutionary optimizers and a standard genetic algorithm for the evolution of digital circuits. The goal was to investigate whether optimizing for circuit speed and area, besides functional quality, can improve the speed of convergence and robustness. We consider robustness a parameter of prime importance, especially for self-optimizing autonomous systems that continuously run the evolutionary optimization process.

Our experimental results show that, for more complex benchmark problems, the classic genetic algorithm is indeed outperformed by two multi-objective optimizers, TSPEA2 and NSGAII. Two further optimizers, SPEA2 and μGA did not perform well for this task. In future, we plan to look at other secondary objectives to improve convergence and robustness. For example, there might be circuit properties besides area and speed that should be enforced. As scalability is one of the main challenges in evolvable hardware, the identification of suitable objectives for a *scalability-driven* evolution is of utmost importance.

Acknowledgment

This work was supported by the German Research Foundation under project number PL 471/1-2 within the priority program *Organic Computing*.

References

1. Sekanina, L.: Evolvable Components. Natural Computing Series. Springer (2004)
2. de Garis, H.: Evolvable Hardware – Genetic Programming of a Darwin Machine. In: Proceedings International Conference on Artificial Neural Networks and Genetic Algorithms (ICAN-NGA), Springer (1993)
3. Higuchi, T., Niwa, T., Tanaka, T., Iba, H., de Garis, H., Furuya, T.: Evolving Hardware with Genetic Learning: A First Step Towards Building a Darwin Machine. In: Proceedings 2nd International Conference on Simulation of Adaptive Behavior (SAB), MIT Press (1993) 417–424
4. Coello Coello, C.A., Aguirre, A.H., Buckles, B.P.: Evolutionary Multiobjective Design of Combinational Logic Circuits. In: Proceedings of the 2nd NASA/DoD Workshop on Evolvable Hardware (EH), Los Alamitos, California, IEEE Computer Society (2000) 161–170
5. Kalganova, T., Miller, J.: Evolving More Efficient Digital Circuits by Allowing Circuit Layout Evolution and Multi-Objective Fitness. In: Proceedings of the 1st NASA/DoD Workshop on Evolvable Hardware (EH), Pasadena, California, IEEE Computer Society (1999) 54–63
6. Kaufmann, P., Platzner, M.: Toward Self-adaptive Embedded Systems: Multi-objective Hardware Evolution. In: Proceedings of the 20th International Conference on Architecture of Computing Systems (ARCS). Volume 4415 of LNCS., Springer (2007) 199–208
7. Miller, J.F., Thomson, P.: Cartesian Genetic Programming. In: Proceedings of the European Conference on Genetic Programming (ECGP), Springer-Verlag (2000) 121–132
8. Coello Coello, C.A., Aguirre, A.H.: Design of Combinational Logic Circuits through an Evolutionary Multiobjective Optimization Approach. In: Artificial Intelligence for Engineering Design, Analysis and Manufacturing. Volume 16., Cambridge University Press (2002) 39–53
9. Zitzler, E., Laumanns, M., Thiele, L.: SPEA2: Improving the Strength Pareto Evolutionary Algorithm. Technical Report 103, Gloriastrasse 35, CH-8092 Zurich, Switzerland (2001)
10. Trefzer, M., Langeheine, J., Meier, K., Schemmel, J.: Operational Amplifiers: An Example for Multi-objective Optimization on an Analog Evolvable Hardware Platform. In: International Conference on Evolvable Systems (ICES), Springer (2005) 86–97

11. Deb, K., Agrawal, S., Pratap, A., Meyarivan, T.: A Fast Elitist Non-dominated Sorting Genetic Algorithm for Multi-objective Optimisation: NSGA-II. In: Proceedings of the 6th International Conference on Parallel Problem Solving from Nature (PPSN), Springer (2000) 849–858
12. Goldberg, D.: Genetic Algorithms in Search, Optimization, and Machine Learning. Addison-Wesley (1989)
13. Coello Coello, C.A., Pulido, G.T.: A Micro-Genetic Algorithm for Multiobjective Optimization. In: First International Conference on Evolutionary Multi-Criterion Optimization (EMO). Volume 1993 of LNCS., Springer (2001) 126–140
14. Knowles, J.D., Corne, D.W.: Approximating the Nondominated Front Using the Pareto Archived Evolution Strategy. In: Evolutionary Computation. Volume 8., MIT Press (2000) 149–172
15. Miller, J.F.: An Empirical Study of the Efficiency of Learning Boolean Functions Using a Cartesian Genetic Programming Approach. In: Proceedings of the Genetic and Evolutionary Computation Conference (GECCO). Volume 2., Morgan Kaufmann (1999) 1135–1142
16. Miller, J.F., Thomson, P., Fogarty, T.: Designing Electronic Circuits Using Evolutionary Algorithms. Arithmetic Circuits: A Case Study. In: Genetic Algorithms and Evolution Strategy in Engineering and Computer Science. John Wiley and Sons (1998) 105–131
17. Kaufmann, P., Platzner, M.: MOVES: A Modular Framework for Hardware Evolution. In: Second NASA/ESA Conference on Adaptive Hardware and Systems (AHS), IEEE (5-8 Aug. 2007) 447–454
18. Koza, J.: Genetic Programming: On the Programming of Computers by Means of Natural Selection. MIT Press (1992)

A Model of Self-Organizing Collaboration

Rumen Andreev

Institute of Computer and Communication Systems, Bulgarian Academy of Sciences, Acad. G. Bonchev str. Bl. 2, 1113 Sofia, Bulgaria, e-mail: rumen@iccs.isdip.bas.bg

Abstract Collaboration joins together persons (active objects) in some activity. The paper concentrates on the activity, since it is the collaboration basis. The basic theory used in computer science for activity analysis is the activity theory that considers activity as a substantial part of the human interaction with the objective reality (environment). In conformance with this theory, the action presents activity substance and operation – activity (action) realization. It is analyzed cooperation as collaboration realized in operation context.

The activity theory considers a kind of activity that does not reveal it as an independent, autonomous thing (self-organizing activity). This paper indicates the main characteristics of self-organizing activity that are basis for content-, context-independent modeling of autonomous activity. The presented model uses formal constructions of the mathematical logic as autonomous frameworks. The self-organizing activity is foundation for modeling of self-organizing collaboration that results in a model describing a collaborative self-organizing system. The latter is framework for a real process and bases on a group of active objects associated by a shared need.

Keywords: Self-organizing activity, cooperative self-organizing system, modeling, reasoning, real process.

1. Introduction

Collaboration indicates a form of participation of persons or active objects in some activity. In accordance with this form, the persons (active objects) are joined together in an activity, i.e. the activity is in state of being joint activity that ensures partnership. The collaboration has two ingredients, since it is a mixture of a group of persons (active objects) and an activity. Our consideration concentrates on the most important ingredient, which is the basis of collaboration - activity. In the computer science, the activity analysis is realized in the light of activity theory.

Please use the following format when citing this chapter:

Andreev, R., 2008, in IFIP International Federation for Information Processing, Volume 268; *Biologically-Inspired Collaborative Computing*; Mike Hinchey, Anastasia Pagnoni, Franz J. Rammig, Hartmut Schmeck; (Boston: Springer), pp. 223–232.

In view of the activity theory, the activity is substantial part of the human inter-action with the objective reality (environment) i.e. the activity is determined through a system that balances [1]. The most fundamental principle of this theory states that the human mind that is one of the main features of human beings is a special component of human interaction with the environment. According to the second principle of this theory, the activity is object-oriented. The object could be not only a material thing of the human environment, but it is considered as a de-sired object (goal), motive or objective, as well. Hence, the activity is a substantial part of a system, which components human beings and the objects of environment interact. The system-relevant analysis of activity is due to its objective character and reveals that the activity takes part in keeping an (eco)system in state of bal-ance. The system is a framework, in which the activity exists.

Another basic principle of the activity theory is the "hierarchical structure of activity" that concerns the activity ontology. The activity exists on three levels with regard to its nature - its existence results of a cause (object). At the highest level the activity is considered as result of motive - object that causes human be-ings to act. On the second level the activity exists as action - goal-directed activity. On the lowest level the actions are realized through operations that are determined by the actual conditions of activity. The operations convert actions into facts. This is a psychologically relevant consideration of activity.

According to a psychologically irrelevant presentation of activity existence, the activity description on the highest level is result of system-relevant analysis that presents its objective character: The activity is a phenomenon of the physical world which existence is explained by the principle of causality. The human that is one of the system components does not cause the activity. It serves for helping bring about the activity, i.e. it is activity factor – agent. As an agent the human be-ing can be both *reactive* and *proactive*. The activity theory adopts the idea that human beings are proactive agents, since the reaction (activity of a reactive agent) is automatic, i.e. unconscious. A proactive agent is able to realize and adopt goals and to take the initiative. It is obviously that on this level the activity analysis and representation can be psychologically irrelevant (objective) while on the other lev-els they are psychologically relevant [2]. The psychologically relevant analysis re-veals the human relevant substance of activity.

The psychologically relevant action is virtual activity, which substance is pre-sented by the line of action. The latter depends on the goal and relates to the way of action happening (course of action). The course fixes the form of action exis-tence (operation form). In view of the physical existence of an action, the activity is considered as operation. When a man operates, we say that he is in action, i.e. he implements its course that serves as purpose of human operation. The purpose causes successful operation (functioning) of human being or executive system. It is considered as objective cause of operation and is well known as *objective*. The operation needs of a plan (organization) of action implementation. Its develop-ment bases on the course of action and resources helping bring about an action. The course of action and resources are two operation factors that determine its context. The former serves as a plan of operation organization. There are two types of resources that guarantee operation: resources that carry out operation (*operation performer*, agent) and resources that support operation performer in doing the action. The purpose (objective) and resources are two interacting factors that are in the base of operation and present an integrated context.

Usually the collaboration is known as cooperation, since it is considered with respect to operation. There are three interacting elements that support the cooperation reality: set of agents, shared resources and shared objective. They determine the framework, which supports an engineer in construction of cooperative forms (Fig. 1). The agents take parts in various cooperative actions, in which they have different roles [3]. They are the substance of cooperation. An agent can either be a human or a computer-based component, which is an active process supported by a computer system [4].

This framework supports the development of various methodologies for cooperation design and realization. The cooperation transforms the set of agents into a group of agents. There are two factors that help the composition of a group: shared resources and a shared view of communicating agents on a subject or shared objective of working agents [5]. The shared resources are contained in environment that is influenced by organizational approaches and coordination techniques. This organized environment is the *background* of cooperation. The shared view is the cooperation *foreground*, i.e. the domain, in which the agents cooperate [6].

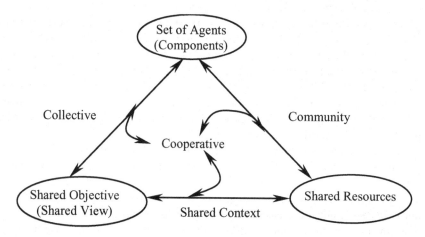

Figure 1. Engineering framework for cooperation

The three cooperation factors act on each other. The shared resources helping transformation of a set of agents into *community* are *community factor*. The community is a cooperative form. The cooperative use of shared resources is a typical community-wide behavior of the members of community, which is considered as a whole. The shared objective is *collective factor*. It helps the union of agents in cooperative form known as *collective*. This is a group of agents that looks like a whole. The shared objective guarantees collective behavior of agents in an organized system. The shared objective and shared resources compose shared context that determines the cooperative operation of agents. They ensure two types of groupware. The group construction is analyzed not only with respect to cooperation, but with regards to communication, as well. The groups regarded in communication assist the cooperation, since the information is a necessary factor for operation management [7]. The shared objective is a factor for organization of resource environment and the shared resources are factor (means) for achievement of a shared objective. The interaction between these cooperation factors ensures an integrated shared context.

The cooperation background can be computer-supported environment (computerized environment), in which the resources are usually distributed. Environment consisted of resources of the same kind is *homogeneous environment* that guarantees uniform access to the resources. If the latter are of different kind, the environment is *heterogeneous*. The unification of the access to the resources of this environment requires its homogenization that can be achieved by using of grid technology [8]. The cooperative use of resources is provided by services for coordination, shared sessions and support of synchronous and asynchronous access [9, 10]. The shared objective that is the logical part of shared context organizes the agents that are workflow participants in a collective system, i.e. coordinates their work [11]. A common project can be regarded as shared objective [12].

The shared view can be used for arrangement of computerized environment, which becomes setting for cooperative work of agents of a group that is regarded as multi-agent system [13]. The shared view is a conceptual model that is factor of interoperability supporting cooperation [14, 15]. In Internet-based environment the common view ensures the building of Semantic Web that ensures cooperation through interoperability [16]. The framework for construction of cooperative forms integrates the main cooperation factors: agents, computerized environment (grid) and shared objective. In this way, it gives an approach to solution of the problem of agents/grid integration [17].

Self-organizing collaboration requires a self-organizing basis, i.e. a self-organizing activity. The self-organizing activity is the substance of autonomous activity, which main characteristic is independence. A thing that is complete in itself (a whole) is an autonomous thing. The self-organizing activity has the form of unity, since it results of arrangement of its parts to form a complete whole. The activity theory explains a kind of activity that is not self-organizing, since it has the following characteristics:

- The activity is substantial part of an integrated system: The integration is due to interaction that ensures system balance. The activity takes part in this interaction, but does not organize it, i.e. it has not unity form;
- As a necessary consequence of an external cause presented by a motive, the activity evidently subjects to the law of causality and its existence *depends on* an external object – the activity is not independent;

- The proactive agent and object that causes the activity are determining elements (*determinants*) of activity and present its context: The *context-dependent* determination of activity is a systematic view on it that does not describes the activity as autonomous thing - independent object with quality of wholeness. It is necessary to distinguish the completeness of an activity from its wholeness that is characteristic of an integrated system.

The next section defines self-organizing activity. On this basis it presents a model of collaborative self-organizing system. The third section describes an engineer view on the content-, context-independent model of self-organizing collaboration.

2. Model of collaborative self-organizing system

The basis of self-organizing collaboration is self-organizing (autonomous) activity. This kind of activity has the following basic characteristics:

- The law of causality is a principle of self-organizing activity: It is in the nature of self-organizing activity
- An autonomous activity results of an intrinsic physical need of an object (human being) and belongs to it, as its capability: As this object holds the activity, it is *active object* that causes an activity and helps for its happening;
- In view of the activity independence, the activity of an active object must ensures the satisfaction of its intrinsic need with the help of another object that must meet this need: This *necessary object* guarantees the activity completion, i.e. it is a part of activity;
- This kind of activity involves in itself all objects that are necessary for its existence, integrates them as parts of a complete whole and arranges them in harmonious relation: The result is the composition of a unity.

These characteristics are necessary for nature-based modeling of self-organizing activity that ensures content-, context- independent representation. The result of this way of modeling is a formal model.

The presented kind of activity materializes the law of causality. Its substance is an integrated system consisting of two interacting components - active object and necessary object. This system guarantees activity wholeness. In self-organizing activity the necessary object does not only correspond to the active object, but corresponds with it, as well. The two objects are in harmonious relation, since the necessary object meets an intrinsic need of the active object. Harmony and equality are simultaneously the most important characteristics of autonomous activity.

According to the nature of self-organizing activity, this kind of activity can be presented by implication, since it present a formal material form that involves two objects that are simultaneously in harmonious relation and in balance. This consideration of implication is in conformance with its definition given by the mathematical logic through a truth table [18]. Using the definition of equivalency, the implication (autonomous activity) has the following presentation, in which a_object is active object and n_object is necessary object

$$\text{a\_object} \Rightarrow \text{n\_object} \equiv (\text{a\_object} \Leftrightarrow \text{n\_object}) \vee (\neg \text{a\_object} \wedge \text{n\_object}).$$

This statement reveals that the implication is union of equivalency and harmonious relation ($\neg$a_object $\wedge$ n_object), where *n_object* satisfies a need of active object ($\neg$a_object). The equivalency *a_object* $\Leftrightarrow$ *n_object* presents an integrated system that bases on the balance due to the interaction between an active object and necessary object. As the integrated system (equivalency) and harmonious relation are the basis of a self-organizing activity, which substance coincides with the definition of unity [19], it is obviously that the implication indicates an autonomous activity, i.e. it is in conformance with unity. Therefore, the formal model of self-organizing activity is *active object* $\Rightarrow$ *necessary object*.

Since the self-organizing activity supports self-organizing collaboration, the model of autonomous activity is a model of the framework for self-organizing collaboration. It represents the outer form of self-organizing collaboration, in which active objects act together. It is accepted that two active objects are partners in a self-organizing activity. This self-organizing collaboration is presented by the following expression

active object$_1$ $\wedge$ active object$_2$ $\Rightarrow$ shared necessary object.

This model of self-organizing collaboration states that the necessary object satisfies a need, which the active object$_1$ shares with the active object$_2$, i.e. it is a shared necessary object (sh_n_object). The self-organizing collaboration is due to *a shared need*.

As result of nature-based (formal) modeling, the model of self-organizing collaboration presents its essential characteristics:

1. Self-organizing collaboration exists in a collaborative self-organizing system. The active objects and necessary object are components of a collective self-organizing system, since they are in symmetrical relation guaranteed by the autonomous activity, in which they take part. The symmetry presented by its qualities of harmony and balance is a characteristic of self-organizing activity.

2. Collaborative self-organizing system is framework for a real process. The self-organizing collaboration has several representations. The upper model is its basic representation that could be presented in the following way

active object1 $\Rightarrow$ (active object2 $\Rightarrow$ shared necessary object).

This expression states that a complex autonomous activity can exists in the   collaborative self-organizing system. The complex autonomous activity is a real process, since it presents a continuous succession of activities. The continuity is a necessary consequence of the wholeness of a object. As the autonomous activity ensures the framework for self-organizing collaboration, the collaborative self-organizing system is a complete whole, as well.

3. Self-organizing collaboration is induced by naturally associated active objects that are in state of being united with an object, which helps them to satisfy their shared need and to integrate them. The following statement presents another alternative expression of self-organizing collaboration

$$\text{a\_object}_1 \wedge \text{a\_object}_2 \Rightarrow \text{sh\_n\_object} \equiv (\neg \text{ a\_object}_1 \vee \neg \text{ a\_object}_1) \vee \text{sh\_n\_object}$$

This expression is inspired by the equivalence between $(p \Rightarrow q)$ and $(\neg p \vee q)$ settled by the mathematical logic. It states that the active objects taking part in collaboration are naturally associated by a shared need, i.e. they are in a group. They are of different types. The shared necessary object integrates them in a whole. They wholeness of the group of active objects is preserved in the framework of self-organizing collaboration.

3 An engineer view on the model of self-organizing collaboration

There is a dualistic view on the representation of self-organizing collaboration that is a special kind of collaboration and is supported and shaped by collaborative self-organizing system. The latter is a material form, in which the collaboration exists. It determines unconditionally an outer form that ensures the happening of self-organizing collaboration. Hence, the collaborative self-organizing system is *a priori form* of self-organizing collaboration and it is a foundation for achievement of a real self-organizing collaboration. In view of an engineer, the presented model is very general description of autonomous collaboration. It is due to the fact that this model does not concentrate on the substance of self-organizing collaboration. The system-relevant analysis of self-organizing activity gives only the outer form (shape), in which it exists and ensures its formal (general) description.

In view of an engineer, the formal description of self-organizing collaboration is useless. An engineer shows interest in the substance of an object that he designs and realizes. For this purpose, it requires a detailed object description that presents the inner form of the object. An inner form that gives the nature of object existence is necessary for object specification. The detailed description does not only ensure a detailed representation of engineer view on object existence, but guarantees object realization, as well.

The satisfaction of engineer requirements can be achieved by a deductive approach to consideration of the model of self-organizing collaboration. It requires more definite presentation of collaboration substance on the levels of action and operation. It has to be in conformance with the limitations of the collaborative self-organizing system. On the level of action it is presented the particular substance of self-organizing collaboration. On the lower level of operation the collaboration presentation is more concrete, since the operation description gives specific knowledge that is necessary for the performance of action.

On the action level, the collaborative self-organizing system has to undergo the following changes: the shared necessary object is treated as shared goal; the active objects are regarded as human beings; the activities of active objects are considered as separate processes, i.e. the whole process is divided in two. To keep the harmony of self-organizing collaboration, the first of the two processes has to produce a satisfactory result, i.e. a result suitable for use of the second process. The second process has to produce a desired result. It is evident that the processes have to be efficient. This is the requirement to the psychologically relevant presentation of self-organizing collaboration on this level. With respect to the operation level, the two main processes are presented as series of actions that are realized through operations. To satisfy the requirement for realization of efficient processes, it is necessary to control the operations. In view of the control, there are the following paradigms of operation performance: *arrangement-conducted operation, operation in regulation* and *adaptation-based operation.*

To be in line with computer science, the self-organizing collaboration could be represented on the action level in the following way

Provider $\wedge$ User $\Rightarrow$ shared goal.

In this model the user and provider are the active objects that are associated by a shared need presented by a shared goal. This model is suitable for representation of an education system, in which the teacher is considered as provider and the student as user. [20]. This statement is useful for representation of a software development system, as well. Here, the software engineer is a provider. The provider can be a producer or servant.

In the following presentation of the upper model of self-organizing collaboration, the provider carries out a producer role

Provider $\Rightarrow$ (User $\Rightarrow$ shared goal).

In this form the model represents the following real meaning (substance) of the analyzed self-organizing collaboration:

- User regulates the producer process and has to achieve a desired result: The producer has to supply the user process with a satisfactory product;
- The provider activity (production process) supports the collaboration and reveals the capabilities of the collaborative self-organizing system: This model of the system is known as process-oriented, since it supports two processes;
- The symmetry of the collaborative self-organizing system is ensured through a harmonious relation between the provider and user: The former has to satisfy the user requirements.

Another form of action-based model of collaborative self-organizing system adopts the idea that the provider is a servant

User $\Rightarrow$ (Provider $\Rightarrow$ shared goal).

This form of the psychologically relevant model of self-organizing collaboration brings out in the foreground, the following essential characteristics of the system:

- The provider is a servant: it is treated as a means in the collaborative process;
- The model presents a service-oriented approach to the realization of collaborative self-organizing system;
- The symmetry is ensured by interaction between the user and provider: In this case the symmetry is presented by its quality of balance;

If the provider is simultaneously a servant and producer, its process has to produce a desired result (not satisfactory result).

4 Conclusions

The analysis of self-organizing collaboration concentrates not on its substance or realization, but on its outer form that coincides with the framework of autonomous collaboration. This approach results in a content-, context-independent representation of collaboration. Since the formal constructions of the mathematical logic are suitable for formal modeling of self- organizing activity and collaboration, the latter are expressed by the five well-known operations (constructions) of the mathematical logic: equivalency, implication, disjunction, conjunction and negation. They are taken up not as structures that serve for presentation of subject substance, but as frameworks (outer form of a whole thing) presenting the material form of self-organizing collaboration.

The different ways of presentation of implication that represent self-organizing activity support our reasoning on self-organizing collaboration. It results in the following findings: self-organizing collaboration exists in the border of an autonomous system; a collaborative self-organizing system is a framework for a real process; naturally associated active objects induce self-organizing collaboration. The usage of the operations of the mathematical logic for presentation of various autonomous frameworks needs of a new interpretation of the basic logic symbols A and $\neg A$ and the basic constructions, in which they take part.

References

1. Kaptelinin, V.: Activity theory: Implications for human computer interactions. In: Brower-Janse, M., Harrington, Th. (eds) Human-Machine Communications for Educational Systems Design, NATO ASI Series F Vol. 129, pp. 5-15. Springer-Verlag (1994)
2. Vicente, K.: Wanted: psychologically relevant, device- and event-independent work analysis techniques. Interact. With Comput. Vol. 11, pp. 237-254 (1999)
3. Tamai, T.: Objects and roles: modeling based on the dualistic view. Inf. and Softw. Technol. Vol. 41, pp. 1005-1010 (1999)
4. Wegner, H., Hupe, P., Matthes, Fl.: A process-oriented and content-based perspective on software components. Inf. Syst. Vol. 25 (2), pp 135-156 (2000)
5. Larsson, T.I., Vainio-Larsson, A.A.: Software producers as software users. In: Gilmore, D.J., Winder, R.L., Detienne, Fr. (eds) User-Centered Requirements for Software Engineering Environments, NATO ASI Series F Vol. 123, pp. 285-306. Springer-Verlag (1994)
6. Lu, St. C-Y, Cai, J.: STARS: A socio-technical framework for integrating design knowledge over the Internet. IEEE Internet Comput. Vol. 4 (5), pp. 54-62 (2000)
7. Guerrero, L.A., Fuller, D.A.: A pattern system for the development of collaborative applications. Inf. and Softw. Technol. Vol. 43, pp. 457-467 (2001)
8. Roure, D., Baker, M.A., Jennings, N.R., Shadbolt, N.R.: The evolution of the grid. In: Fox, G., Hey, A.J.G. (eds) Grid Computing-Making the Global Infrastructure a Reality pp. 65-100, John Wiley and Sons (2003)
9. Lopez, P.G., Skarmeta, A.F.G.: ANTS framework for cooperative work environments. IEEE Comput. Vol. 36 (3), pp. 56- 62 (2003)
10. Singh, M.P.: Conceptual modeling for multiagent systems: Applying interaction-oriented programming. In: Chen, P.P., Akoka, J. (eds) Conceptual Modeling, LNCS Vol. 1565, pp.195-210, Springer-Verlag (1999)
11. Gellersen, H.-W.: Support of user interface design aspects in a framework for distributed cooperative applications. In: Taylor, R.N., Coutaz, J. (eds) Software Engineering and Human-Computer Interaction, LNCS Vol. 896, pp. 196-210, Springer-Verlag (1995)
12. Perez, M., Rojas, T.: Evaluation of workflow-type software products: a case study. Inf. and Softw. Technol. Vol. 42, pp. 489-503 (2000)
13. Liedekerke, M.H., Avouris, N.M.: Debugging multi-agent systems. Inf. and Softw. Technol. Vol. 37 (2), pp. 103-112 (1995)
14. Szykman, S., Fenves, St.J., Keirouz, W., Shooter, St.B.: A foundation for interoperability in next-generation product development systems. Comput.-Aided Des. Vol. 33, pp. 545-559 (2001)
15. Chen, P.P., Thalheim, B., Wong, L.Y.: Future directions of conceptual modeling. In: Chen, P.P., Akoka, J. (eds) Coceptual Modeling, LNCS Vol. 1565, pp. 287-301, Springer-Verlag (1999)
16. Aroyo, L., Dolog, P., Houben, G.-J., Kravcik, M., Naeve, A., Nilsson, M., Wild, Fr.: Interoperability in personalized adaptive learning. Educ. Technol. and Soc. Vol. 9 (2), pp. 4-18 (2006)
17. Foster, I., Jennings, N.R., Kesselman, C.: Brain meets brawn: Why grid and agents need each other. In: Proceedings of AAMAS'04, ACM (2004)
18. Stall, R.R.: Sets, Logic and Axiomatic Theories. Freemand & Company, London (1965)
19 Hornby, A.S., Cowie, A.P., Gimson, A.C.: Oxford Advanced Learner's Dictionary of Current English (Third Edition). Oxford University Press, Oxford (1987)
20. Andreev, R.D., Troyanova, N.V.: E-learning design: An integrated agent-grid service architecture. In: Proceedings of IEEE John Vincent Atanasoff 2006 International Symposium on Modern Computing, pp. 208-213, IEEE Computer Society (2006)

Guiding Exploration by Combining Individual Learning and Imitation in Societies of Autonomous Robots

Willi Richert, Oliver Niehörster, Florian Klompmaker

Abstract Robots have a powerful means to drastically cut down the exploration space with imitation. However, as existing imitation approaches usually require repetitive demonstrations of the skill to learn in order to be useful, those are typically not applicable in groups of robots. In these settings usually each robot has its own task to accomplish and should not be disturbed by teaching others. As a result an imitating robot most of the time has only one observation of a specific skill from which it can learn.

We present an approach that allows an individually learning robot to make use of such cases of sporadic imitation which is the normal case in groups of robots. Thereby, a robot can use imitation in order to guide its exploration efforts towards more rewarding areas in the exploration space. This is inspired by imitation often found in nature where animals or humans try to map observations into their own capability space. We show the feasibility by realistic simulation of Pioneer robots.

1 Introduction

With the benefits of drastically cutting down the exploration space imitation is one of the most powerful learning techniques one can find in nature [5, 6, 8]. This has been acknowledged also by robotics researchers when they embraced different methods to apply imitation to learn tennis swings or drumming movements [10] or e.g. to forage [9]. However, except for the work on imitating skill sequences in all these experiments the demonstrator is always determined (often the human) and the time frame where the imitation has to pay attention is provided beforehand. The task to be learned by imitation is then repeated several times and the robot afterwards has to derive a generalized representation of the imitated task and be able to replay it. Up to now no research has been carried out regarding sporadic imitation, which is apparently very important when robots in groups should benefit from each

Intelligent Mobile Systems, University of Paderborn / C-LAB, Germany, richert@c-lab.de

Please use the following format when citing this chapter:

Richert, W., Niehörster, O. and Klompmaker, F., 2008, in IFIP International Federation for Information Processing, Volume 268; *Biologically-Inspired Collaborative Computing*; Mike Hinchey, Anastasia Pagnoni, Franz J. Rammig, Hartmut Schmeck; (Boston: Springer), pp. 233–244.

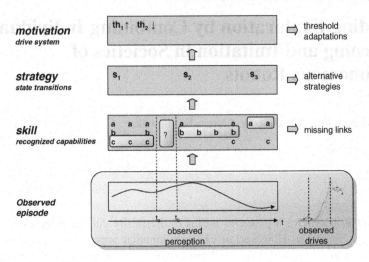

Fig. 1 Procedure of interpreting another robot's performance in order to imitate it.

others learning efforts. Typically, the imitation process should not interrupt the observed robot, so that the imitating robot often has only one example of the same type of interesting behavior to learn from. As this usually does not provide enough information for learning a generalized version of the observed action, it can help the observer to narrow the learning exploration space. This is the aim of our paper.

With the presented approach comprising the strategy and low-level skill layers an observing robot can benefit from the imitation process

1. by observing new state sequences for which it could spent more exploration efforts,
2. by observing new behaviors for already known state transitions, and
3. by incorporating other robot's transition data condensed into its own strategy.

In Fig. 1 an example is shown in which the robot (imitator) tries to *understand* the observed behavior episode of another robot (demonstrator). The observed episode consists of the recorded perception and the demonstrator's visible "well-being", a kind of emotional state that comprises its overall state in form of a set of drives. Therefore the imitator first translates the observations into its own perception to see what it would perceive if itself would have been in the demonstrator's situation. It then scans the subjective perception and allows its low-level skill to give so-called *votes* about how well each skill could have achieved the perception changes. Using an algorithm inspired by Viterbi those votes are then used together with the likeliness of the demonstrator's state space to find the most likely path corresponding to the observations. In this paper we will focus at the skill and strategy layer, as they are most important to the understanding of the observed behavior.

2 Related Work

Most approaches regarding imitation of robotic behavior are based on Hidden Markov Models (HMM) and use the Viterbi algorithm to synthesize behavior thereof. Billard et al. [4] use e.g. the Viterbi algorithm to let the upper part of a robot replay a limited set of arm movements that move colored objects. In their work the demonstrator-imitator roles are known and fixed. Also the start and end points of the behavior to imitate is known to the robot. They split the imitation task into the observation and imitation processes, having the goal to minimize the discrepancy between the demonstrated and imitated data sets. In their approach the robot is only able to learn low-level behavior and this can only be done from scratch. In contrast to Billard we do not aim to imitate for the sake of copying another robot's low-level behavior, but to gather new inspiration for the imitating robot to drive its learning efforts to. This will have to include all levels of abstraction, not only low-level behavior.

Closest to our approach come Inamura et al. [11, 12] with their *Mimesis Loop* approach. Thereby they are able to symbolize observed low-level behavior traces. This is used as top-down teaching from the user's side in combination with the bottom-up learning from the robot's side. As this is useful to decrease the programming effort it is an exclusive solution, not allowing to be used with other learning techniques like e.g. Reinforcement Learning. Also their approach is not able to use already existing abstract states of the imitator in the recognition process. Once a robot has extracted enough information to construct a HMM based on the recognized low-level behaviors it is fixed to that HMM – no exploratory actions on the abstract states are possible any more. Furthermore, the segmentation process that splits the continuous movement trajectories into basic movements uses a fixed scheme. With that it is not possible to allow for ambiguities at the recognition phase.

In our approach we assume that the robot has already decent self-learning capabilities. Imitation is used to guide the robot to the "salient" points in exploration space. With more experience the robot will have collected better skills and a more realistic strategy representation. This in turn will enable it to extract more knowledge from its observation efforts.

3 Imitation Supporting Architecture

The desired outcome of the observation and recognition phase in an imitation process is a state-action-trace that results in a performance similar to the observations. For this the robot has to find abstract states in its own strategy that should play a role when replaying the imitated behavior. Furthermore, it should only regard states that can be connected via actions the imitating robot is capable of. This leads to a tight coupling of the strategy and skill component in the system architecture, accomplishing the recognition of other robots in terms of its own strategy and skill capabilities. Therefore, the strategy and skill layers will be described before. The strategy is modelled with a Semi-Markov Decision Process (SMDP) that has a dynamically adjustable state space (Sec. 3.1) and uses self-developed skills as actions, which are triggered in terms of goal functions on the perception (Sec. 3.2).

3.1 Learning strategies

The strategy layer is inspired by the AMPS approach [13]. It uses a domain-dependent abstraction method to generalize actual state realizations into abstract regions (in our work we use nearest neighbor [7]). The Reinforcement Learning algorithm is then applied onto these regions which simplifies and speeds up the whole process significantly. As the regions can be merged and split at run-time we use Value Iteration [15] to determine the best policy. AMPS, however, applies the splitting and merging also to the action space, which works fine in artificial domains but will not cope with the domain dependency one is typically faced with in real environments. Here, we use as the strategy's actions goal functions which have to be realized by a separate skill learning layer.

In contrast to the pure AMPS method, which by the nature of Reinforcement Learning always learns one strategy to reach one goal, self-adapting systems often have to fulfill several goals – sometimes contradictory ones. Take for example a system that has to fulfill a task while paying attention to its diminishing resources. If it accomplishes the task the resources might get exhausted. On the other hand, if it always stays near the fuel station, the task won't be accomplished. As already described we use abstract drives which the designer has to specify. These drives may also contain competing goals. The big advantage of our approach is that the robot can learn a separate strategy for each drive. Depending on how big the actual motivation is for every drive it has now a means to choose the right strategy for the actual perception and drive state.

3.2 Learning low-level skills

The input of the skill-learning algorithm is given by the strategy layer (Sec. 3.1) in terms of an error function e. Fig. 3 shows a camera image that has been taken from the robot used in the experiments (cf. Sec. 5). The ball that is recognized by a vision algorithm has the properties width and the 2D coordinates of the image. Let d be the euclidean distance of the ball to the image center and Δr the difference between the maximum size of the vision image and the size of the ball in it. The error function that formulates the goal to maximize the ball in the middle of the camera image e.g. would be:

$$e(d, \Delta r) = \sqrt{d^2 + \Delta r^2} \qquad (1)$$

The first step is then to get a set of training examples that will later be generalized. During this initial exploration the algorithms gathers information about the relationship between the actuators and the effects. The changing of the actuators is called an *action A*. An *effect* is the perceived result of an action. The actions are generated randomly and are applied for some time. In this phase we call the actuator values the input I and the perceived effect the output O, as they are seen from the skill learning algorithm's perspective. This information are the components of a trace T with the length t: $T = (A, \{(I_0, O_0), \ldots, (I_{t-1}, O_{t-1})\})$. Several traces are recorded. Now the error function e is used to extract the good traces forming the training set. To get as many traces as possible, every trace is cut at the position i $(0 \leq i \leq t-1)$ of the lowest error $e(O_i)$. Then every trace not leading to an improvement in terms of e is discarded.

Previous to the generalization, the number of traces and the dimensions have to be refined to reduce the generalization complexity. The most important attributes of a trace are A, I_0, O_0 and O_{t-1}. If the actuator configuration I_0 and the sensor vector O_0, which describe the current situation, are given, A has to be used to reach the effect O_{t-1}. To reduce the number of traces, we do an agglomerative hierarchical clustering. Only the mentioned attributes of a trace are used. The distance measure between two traces is the euclidean distance of the attribute values. The distance between two clusters is defined by average-linkage. The dimensions of I_0 and O_0 depend on the number of actuators and effect properties. Actuators that don't influence the effect can be ignored in the generalization step. Another side-effect of the dimension reduction is the noise reduction, because also the data dimensions with no significant effect to the action-effect can be ignored. We use PCA [1] for this and specify the number of principal components to be kept by the fraction of variance to be explained. In our experiments we were able to reduce the dimensions from six to two while maintaining 95 percent of the data's accuracy.

The last two steps reduce the trace data to the basic properties. In the PCA step a mapping from the data into a new artificial space is done. To generalize the data, a mapping from the principal components. $x_0, \ldots, x_n$ to the individual actor elements $a_i \in A$ is calculated. We use a polynomial regression for every a_i. To get the simplest polynomial of $\sum_{i=0}^{d} \prod_{j=0}^{n} p_{ij} x_j{}^i$ that fits the data sufficiently the algorithm starts with $d = 0$ and increments it until the prediction error drops below a predefined error threshold. This process can be seen in Fig. 2. There is also a threshold for the complexity's degree. To avoid over-fitting a maximal possible degree can be specified. Finally, a function $f_i(x_0, \ldots, x_n) = a_i$ is calculated for each a_i. When applying the learned skill I_0 and O_0 are known as the current parameter values of the actuators and sensors. A mapping to the PCA space has then to be done before using the calculated f_is to build the next action A. With this approach the robot can reach maximal adaptivity and robustness with regard to sudden breaks or graceful degradation [14].

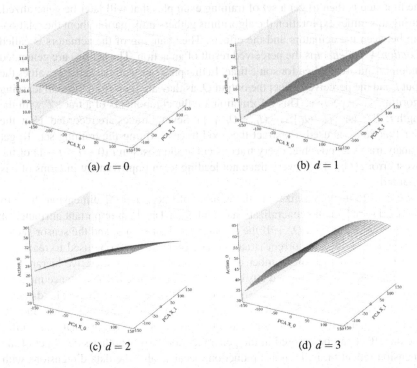

(a) $d = 0$ (b) $d = 1$

(c) $d = 2$ (d) $d = 3$

Fig. 2 Finding the simplest reasonable hypothesis for the first actor element in PCA space. The graphs show the fitted function for one actor dimension dependent on the two calculated PCA dimensions. The degree of the polynomial is incremented from $d = 0$ up to $d = 3$. In Fig. 2(d) the final function can be seen. The increase of d has been stopped because the fitting error falls below a defined threshold.

4 Sporadic Imitation

With the described means for strategy and skill learning we can now adapt the Viterbi algorithm which is often used to imitate using HMMs. Before we explain the core of our imitation algorithm, we will therefor give a short overview of the Viterbi algorithm following the notation of Bengio [3].

4.1 Viterbi

The Viterbi algorithm [16] tries to find the most likely hidden state sequence s_1^T (*Viterbi path*) that explains the observation sequence o_1^T. This can be done by maximizing the following constraints:

$$s_1^{T*} = \arg\max_{s_1^T} P(s_1^T \mid o_1^T) = \arg\max_{s_1^T} P(s_1^T, o_1^T) \qquad (2)$$

Using Bellman's dynamic programming algorithm [2] the Viterbi algorithm determines the maximum efficiently in time $O(Tn)$ where n is the number of non-zero transition probabilities. It recursively calculates the probability

$$V(s,t) = \max_{s_1^{t-1}} P(o_1^t, s_1^{t-1}, s_t = s) \qquad (3)$$

that s is the hidden state at time t given the observations o_1^t for all $s \in S$:

$$V(s,t) = P(o_t \mid s_t = s) \max_{s'} P(s_t = s \mid s_{t-1} = s') V(s', t-1) \qquad (4)$$

V is initialised with $V(s,1) = \max_{s_1} P(o_1 \mid s_1 = s) P(s_1 = s) \; \forall \, s \in S$. The most likely path can now be extracted using

$$\varphi(s,t) = \arg\max_{s'} P(s_t = s \mid s_{t-1} = s') V(s', t-1), \qquad (5)$$

which determines the best predecessor of state s at time t.

4.2 Understanding observed behavior

The imitation approaches usually found in literature calculate the Viterbi path to find the state sequence the imitator should realize in order to exactly copy the observed behavior. This is done using the state space (assumed to be fix) of the inferred HMM, which is assumed to reflect the demonstrator's state space. In contrast to those methods it is important to see that we use a method similar to the calculated Viterbi path to explain the observations recorded from the demonstrator with the imitator's already existing state and action space. Thereby, the imitator tries to *understand* the demonstrator with the knowledge it already has in terms of its own state space (cf. Sec. 3.1) and behavior repertoire (cf. Sec. 3.2).

If the observations provide enough information to infer the corresponding state, $P(o_t \mid s_t)$ could be straightforwardly calculated out of the state representation chosen for the specific domain. If e.g. a nearest neighbor approach is chosen to map state observations to abstract states used in the SMDP, $P(o_t \mid s_t)$ could e.g. chosen to be inversely dependent on the distance to the labeled observation instances in the kNN-representation. However, this is seldom the case in realistic applications so that in order to be able to use Viterbi for inference on the imitator's self-learned knowledge, the robot has to 1) infer the probable state transitions, and 2) guess which of its behaviors could have realized those observed state transition.

The calculation of $P(s_t = s \mid s_{t-1} = s')$ in Eq. 4 is more involved. If one would just take the transition probability of its greedy action in s_{t-1} the robot would not get new insight about other and maybe better state transition behaviors in that specific state. Instead, it should guess from the observations which of the behavior in its own behavior repertoire would best match the recorded observations.

Let us now consider state transition $\langle s_{t_a}, s_{t_b} \rangle$, where $s_{t_a} \neq s_{t_b}$. Firstly, for every recorded observation step $\langle o_{t-1}, o_t \rangle$ ($t \in [t_a, t_b]$) all the behaviors are asked to give a vote $P_b(o_t | o_{t-1})$ representing the ability of behavior b to be able to realize that step[1]. These are determined by means of the corresponding error function with which the behaviors were learnt. These votes are then divided by the time span of the full state transition:

$$P_b(s_{t_b} | s_{t_a}) = \frac{\sum_{t=t_a}^{t_b} P_b(o_t | o_{t-1}, s_{t_a})}{t_b - t_a} \tag{6}$$

At every state transition, one can now determine the most likely transition action $b_{ml} = \arg\max_b P_b(s_{t_b} | s_{t_a})$. It can be used to retrieve the transition probability in the observer's SMDP that would most probably correspond to the observation of the demonstrator: $P(s_{t_b} | s_{t_a}) = P(s_{t_b} | s_{t_a}, b_{ml})$. Thereby, we get the recursive solution

$$V(s,t) = \max_b P_b(o_t | s_t = s, o_{t-1}) \max_{s'} P(s_t = s | s_{t-1} = s', a = b_{t-1}) V(s', t-1), \tag{7}$$

in which $P(s_t = s | s_{t-1} = s', a = b_{t-1}) = T(s', a, s)$ are the transition probabilities learnt in the strategy layer. $\varphi(s,t)$ is determined accordingly. For full reference, the whole algorithm is depicted in Alg. 1. It has the same time complexity as the Viterbi algorithm.

With this information the observing robot can now either remember the $\langle s_i, a, s_{i+1} \rangle$-traces for later replay or spend direct reward along that trace in its strategy layer. If $P(s_{t_b} | s_{t_a}, a_{greedy})$ is below a predefined threshold (θ in Alg. 1) it is assumed that the robot has no behavior that could probably generate the observed movement from time t_a to t_b, marking where it could most efficiently spend its valuable exploration time. Of course, in this case it is wise not to incorporate the understood sub-sequences of the observed trace, but to wait until behavior for the missing link has been learnt so that the full trace is understood.

5 Evaluation

To evaluate the approach robots were put into an environment with soccer balls that had to be transported onto an elevated platform (Fig. 4). To achieve that they can simply push the ball or use their grippers to pick the ball and release it on the target area. The robots have a defined field of view (fov) of $60°$. The field size is $100m^2$. They are able to perceive via their vision capabilities the distance and bearing of the nearest soccer ball if it is within their fov. The platform onto which the ball has to be put is given as absolute coordinates to the robot, which also knows its own position. Overall the robot can perceive the following attributes:

---

[1] Note the different time scales at the observation and state recordings notations.

Algorithm 1 RECOGNIZE: Recognize familiar behavior and save unrecognizable behavior for later exploration

---

Input: O_1^T: observation $\langle (o_1, e_1), \dots, (o_T, e_T) \rangle$ as an (observation, evaluation)-episode stream where $e_1 < e_T$; S and $T(s', a, s)$: state space and transition probabilities of the SMDP

Output: Recognized most likely state transitions and missing links

1: Transform O_1^T into subjective observations $\rightarrow o_1^T$
2: $\Gamma \leftarrow \emptyset$ // collects *understood* $\langle s', a, s \rangle$ triples
3: $\Psi \leftarrow \emptyset$ // collects missing links $\langle s', s \rangle$ that must be explored later on
4: $V(s, 1) \leftarrow \max_{s_1} P(o_1 \mid s_1 = s) P(s_1 = s) \; \forall s \in S$
5: $t_{last} \leftarrow 1$
6: $t \leftarrow 2$
7: **while** $t < |T|$ **do**
8:     **for** $s \in S$ **do**
9:        $b_{t-1} \leftarrow \arg\max_b P_b(o_t \mid s_{t-1} = s, o_{t-1})$
10:        $V(s, t) \leftarrow \max_b P_b(o_t \mid s_{t-1} = s, o_{t-1}) \max_{s'} T(s', b_{t-1}, s) V(s', t-1)$
11:        $\varphi(s, t) \leftarrow \arg\max_{s'} T(s', b_{t-1}, s) V(s', t-1)$
12:        **if** $\varphi(s, t) \neq \varphi(s, t_{last})$ **then**
13:           $s'_{last} \leftarrow \varphi(s, t_{last})$
14:           $b_{ml} \leftarrow \arg\max_b P_b(s \mid s_{t_{last}})$
15:           $s_{last} \leftarrow \varphi(s, t)$
16:           $\Gamma \leftarrow \Gamma \cup \langle s'_{last}, b_{ml}, s_{last} \rangle$
17:           $t_{last} \leftarrow t$
18:           **break**
19:        **end if**
20:     **end for**
21:     **if** $\max_b P_b(o_t \mid s_{t-1}, o_{t-1}) < \theta$ **then**
22:        **while** $\max_b P_b(o_t \mid o_{t-1}) < \theta$ **and** $t < |T| - 1$ **do**
23:           $t \leftarrow t + 1$
24:        **end while**
25:        $\Psi \leftarrow \Psi \cup \langle t_{last}, t \rangle$
26:        $V(s, t) \leftarrow \max_{s_t} P(o_t \mid s_t) P(s_t) \; \forall s \in S$
27:     **end if**
28:     $t \leftarrow t + 1$
29: **end while**
30: **return** Γ, Ψ

- the relative coordinates and width of the ball in the camera image (Fig. 3): (x_b, y_b, w_b)
- distance of the ball from the ground the robot is standing on to detect gripper activities: h_b
- distance and bearing of the target zone: (d_z, θ_z)

The skill layer has as the capabilities to rotate and to translate the robot and to manipulate the gripper. If it is not using its grippers it is nevertheless able to move the balls around in the field by simply pushing them. In the imitation process only the positions of the ball and the robots can be observed. The robot's perception, called *observation* in the algorithm, is thus:

$$o = (x_b, y_b, w_b, h_b, d_z, \theta_z)$$

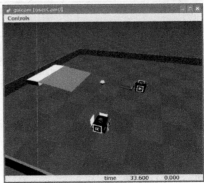

Fig. 3 The perception of the robot that we use in our experiments. It shows one camera image with the ball that has been recognized by our vision algorithm.

Fig. 4 Experimental scenario: The ball has to be put onto the elevated platform.

Changes in this data are used by the already learned skills in the recognition process to check whether the they could have accomplished those changes.

The experiment goes as follows: Two robots, called demonstrator and imitator in the following, are equipped with appropriate strategies and skills in order to move the ball around: The demonstrator is setup with manually handcrafted code, comprising three skills: approaching ball, lifting the ball, and approaching the goal. The imitator is missing the behavior to lift the ball. It has instead individually learned the skill to approach the ball (cf. Sec. 3.2) and is able to approach the goal, and the corresponding strategy using those skills (cf. Sec. 3.1). In the experiment, the imitator is allowed to observe the demonstrator. The new exploration hints as received from the presented algorithm it has obtained via the observation process are then analyzed. The actual exploration in thereby collected narrowed exploration space is not focused in this paper.

As can be seen in Fig. 5 the imitator has successfully recognized episodes in the demonstrator's movements that coincide with the imitator's own behavior knowledge (dark areas). The B denotes the time span in which the demonstrator recognized a behavior resembling its own *approach ball* behavior, and G resembling its *approach goal* behavior. It is interesting that the imitator even was able to detect *not understandable* behavior as such (light areas) and bootstrap the recognition process as soon as it has reasonable explanations for the observed behavior data. This *missing link* can now be used in the subsequent exploration processes to direct the exploration towards it, while the understandable regions can be used, e.g., to adapt the strategy towards more aggressively using them.

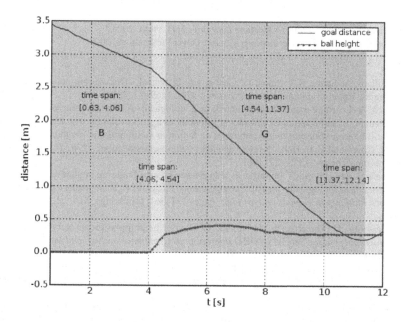

Fig. 5 Recognition results during the imitation process: *B* and *G* (dark area) denote the behavior in time that the demonstrator has understood as equivalent to its *approaching ball* and *approaching goal* behavior. The behavior between them (light area), lifting the ball, is recognized as a missing link.

6 Conclusion

We have shown how sporadic imitation can be accomplished to guide the exploration efforts towards more interesting spaces. For the first time it was shown how inspired by the Viterbi algorithm the maximum likely path of states can be found corresponding to the observation with full reference to the observers own already learned low-level skill capabilities. With it, the observer could reliably explain the demonstrator's performance in terms of its own capabilities if it had skills that could describe the observations or recognize intervals in the observation that could not be understood and should be explored in more detail later on.

Future research should concentrate on more fine-grained dissemination of the *unknown* regions. Using $P_b(s_{t_b} | s_{t_a})$ (Eq. 6) the robot is not able e.g. to detect that more than one action is necessary to be explored in order to accomplish the state transition $\langle s_{t_a}, s_{t_b} \rangle$. Here it would be helpful to look for consecutive ε-homogeneous action sequences. Such a sequence would then contain actions of the same type with probability $1 - \varepsilon$.

References

1. E. Alpaydin. *Introduction To Machine Learning*. MIT Press, 2004.
2. R. Bellman. *Dynamic Programming*. Courier Dover Publications, 2003.
3. Y. Bengio. Markovian models for sequential data. *Neural Computing Surveys*, 2:129–162, 1999.
4. A. Billard, Y. Epars, S. Calinon, G. Cheng, and S. Schaal. Discovering optimal imitation strategies. *Robotics and Autonomous Systems*, 47(2-3):69–77, 2004.
5. Billard, A. Learning motor skills by imitation: a biologically inspired robotic model, 2000.
6. Borenstein, E. and Ruppin, E. Enhancing autonomous agents evolution with learning by imitation. In *Second International Symposium on Imitation in Animals and Artifacts*, 2003.
7. T. M. Cover and P. E. Hart. Nearest neighbor pattern classification. *IEEE Transactions on Information Theory*, 13:21–27, 1967.
8. Demiris, J. and Hayes, G. Imitation as a dual-route process featuring predictive and learning components: a biologically-plausible computational model. In K. Dautenhahn and C. Nehaniv, editors, *Imitation in animals and artifacts*, pages 327–361, Cambridge, MA, USA, 2002. MIT Press.
9. Y. Gatsoulis, G. Maistros, Y. Marom, and G. Hayes. Learning to forage through imitation. In *Proceedings of the Second IASTED International Conference on Artificial Intelligence and Applications (AIA2002)*, pages 485–491, Sept. 2002.
10. A. Ijspeert, J. Nakanishi, and S. Schaal. Movement imitation with nonlinear dynamical systems in humanoid robots. In *International Conference on Robotics and Automation (ICRA2002)*, 2002.
11. T. Inamura, Y. Nakamura, H. Ezaki, and I. Toshima. Imitation and primitive symbol acquisition of humanoids by the integrated mimesis loop. *Robotics and Automation, 2001. Proceedings 2001 ICRA. IEEE International Conference on*, 4, 2001.
12. T. Inamura, I. Toshima, Y. Nakamura, and J. Saitama. Acquiring Motion Elements for Bidirectional Computation of Motion Recognition and Generation. *Experimental Robotics VIII*, 2003.
13. M. J. Kochenderfer. *Adaptive Modelling and Planning for Learning Intelligent Behaviour*. PhD thesis, School of Informatics, University of Edinburgh, 2006.
14. W. Richert, O. Lüke, B. Nordmeyer, and B. Kleinjohann. Increasing the autonomy of mobile robots by on-line learning simultaneously at different levels of abstraction. In *IEEE International Conference on Autonomic and Autonomous Systems (ICAS'08)*, March 2008.
15. R. S. Sutton and A. G. Barto. *Reinforcement Learning: An Introduction*. MIT Press, Cambridge, 1998.
16. A. Viterbi. Error bounds for convolutional codes and an asymptotically optimum decoding algorithm. *Information Theory, IEEE Transactions on*, 13(2):260–269, 1967.

Author Index